VEHICLE HANDLING DYNAMICS

Vehicle Handling Dynamics

by

J.R.Ellis

Mechanical Engineering Publications Limited
LONDON

First published 1994

ISBN 0 85298 885 0

A CIP catalogue record for this book is available from the British Library.

Printed in the United Kingdom by Page Bros, Norwich

CONTENTS

FOREWORD

My original text on vehicle handling, *Vehicle Dynamics* (Business Books) was published in the mid 1960s, It has now been out of print for a number of years, but during that time there has been a steady stream of requests for copies, This present text covers essentially the same subject area as the earlier book, but modified in the light of its use as an advanced teaching text, in the light of feedback from students at the Cranfield Institute of Technology, and in the light of consulting and lecturing activities with tyre and vehicle manufacturers.

Tyre models are now usually numerical curve fits of test data, with some attempts to apply structural techniques based on shell theory. Thus, a typical curve fitting routine is given, although Bezier or spline routines may be used with equal success. The 'taut string' model of 1942, developed by Temple and Von Schlippe, to explain wheel shimmy has been deleted, and a lateral spring model from which a first order differential equation is generated is given.

The roll centre concept for graphic analysis of static suspension qualities remains a useful basic design tool, and I am frequently delighted by its simplicity. It was never intended to describe the dynamic responses of a vehicle by postulating that roll occurred around an axis defined by joining roll centres for front and rear axes, since this suggests that the tyre forces do not affect the vehicle roll mode! The new handling model uses principle axes of inertia and suspension derivatives to generate attitude angles for tyre force and moment, and provides a more realistic view of tyre performance.

Many of the results included in the book were obtained by mechanising the differential equation to which they refer in BASIC or FORTRAN. Compiled BASIC is a good option for use on a PC since graphs can be created easily. A fourth order Runge-Kutta integration routine was used.

John Ellis
September 1993

CHAPTER ONE

The Pneumatic Tyre

From the time of its initial conception about 1877 the pneumatic tyre has been the subject of continuing development, firstly to satisfy the needs of bicycle comfort and later to meet the more stringent requirements of the automotive and aircraft industries. This chapter describes those characteristics of tyres which are of interest and application in the field of vehicle dynamics. The subjects discussed are as follows:

- Mechanical properties of the static tyre.
- Friction between tyre and road.
- The tyre axis system.
- Development of lateral force and steering moment (F&M)
- Models for F&M.
- Transient steering models.
- Tyre terrain envelopment models.
- Tyre vibrations.

1.1 MECHANICAL PROPERTIES OF TYRES

The spring-like responses of the wheel to forces applied in any of the three

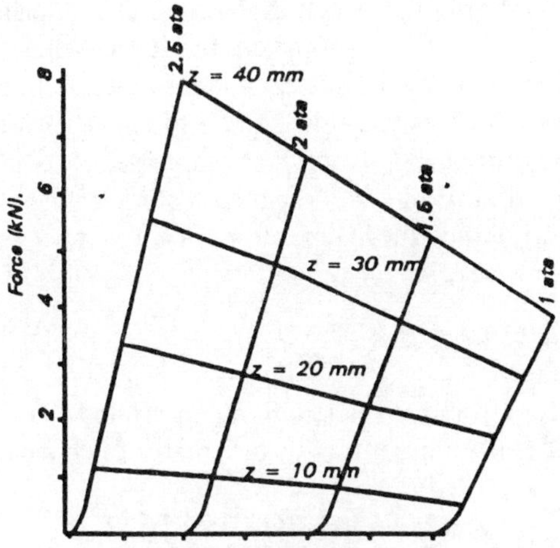

Fig.1.1(a) The normal stiffness of a tyre

directions, x, y, z, are used in many vehicle models.

Figure 1.1. is an example of data from static tests. The normal force vs deflection shows a small toe in which the initial flattening of the tyre contact length occurs; after this the force/deflection characteristic is linear. This graph is presented as a 'carpet' in which the results of testing at a series of inflation pressures are connected by shifting the origin of each full line to the right by an amount proportional to the change in inflation pressure.

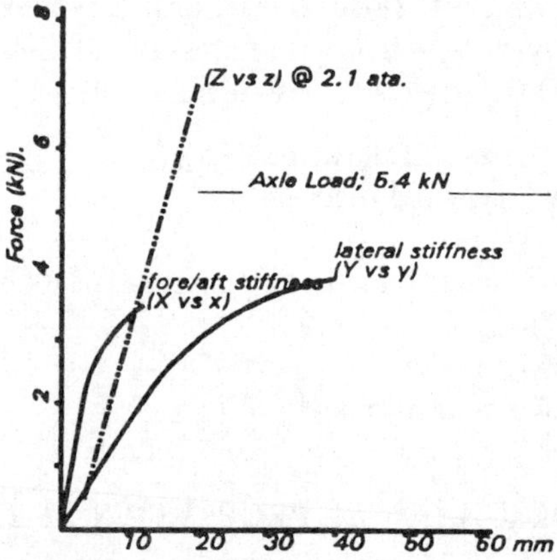

Fig.1.1(b) Fore/aft and lateral tyre stiffness

Lateral and longitudinal force/deflection characteristics are measured by applying a force in the appropriate direction. When the tyre is already loaded against the road by a force, Z, these forces are limited by the sliding of the tyre. The initial response of the tyre to lateral or longitudinal force in the road plane is linear, but as the force increases parts of the contact patch creep on the road until the whole patch is sliding when the applied force reaches its maximum value. Generally, a tyre is most stiff in the fore/aft direction and least stiff laterally.

$$K_x > K_z > K_y$$

The static tyre can also resist a moment around the z axis by virtue of the distribution of the normal force over an area of contact.

1.2 TYRE/ROAD FRICTION

When a tyre contacts the road surface the soft rubber of the tread drapes itself around the hard asperities of the road, most of the deflection occurs in the tyre and the road is relatively unchanged. The intrusion of the road surface into the tyre produces a different contact condition from that which is experienced when two metallic surfaces are brought together, and the concept of friction between tyre and road reflects that difference. For dry metallic surfaces in contact a 'stick-slip' process in which local high spots are continuously pressure welded together and then torn apart as the surfaces slide is an accepted description of friction. In rubber there is the additional factor of the movement of the high points on the road through the soft rubber to be explained and the total resistance to sliding will have components due to the typical friction process and the work done by that movement.

Each penetration of the tyre by the road produces a high local force assumed to be reacted by a hydrostatic force distribution within the rubber. Figure 1.2(a) shows the static loading condition. Sliding the tyre on the

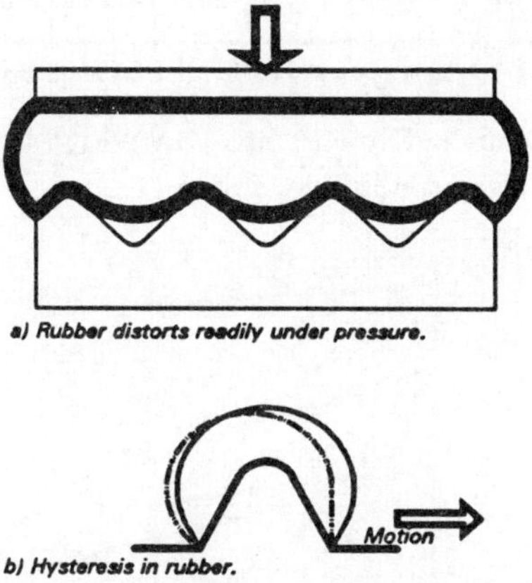

Fig.1.2 The friction of rubber depends upon the macro surface of the road

road causes these local force high spots to move through the rubber, the loading/unloading cycle absorbs energy due to hysteresis, and thus a force is required to sustain motion even in the absence of mechanical friction. This effect is shown as a skewness in the pressure diagram in Fig. 1.2(b).

Surface temperature also affects the sliding resistance. Above freezing there is a steady decline with temperature fall, but around freezing there is a sudden fall in resistance followed by a rise to a value somewhat below that previously attained. Entrapped moisture is the apparent cause of the discontinuity around freezing, while the general reduction with temperature is probably due to hardening of the rubber as the temperature falls.

Road surfaces may be classified by a combination of the surface texture on a macro and micro scale. Figure 1.3 shows such a classification. The micro scale harshness is effective in maintaining tyre steering and traction properties on wet surfaces without standing water.

SURFACE APPEARANCE	TEXTURE	
	MACRO	MICRO
A	HARSH	HARSH
B	HARSH	POLISHED
C	SMOOTH	HARSH
D	SMOOTH	POLISHED

Fig.1.3 A visual classification of road surfaces

Speed reduces friction coefficients in a way which is dependent upon both road surface and surface condition. Figure 1.4. shows the effects of speed for two wet roads.

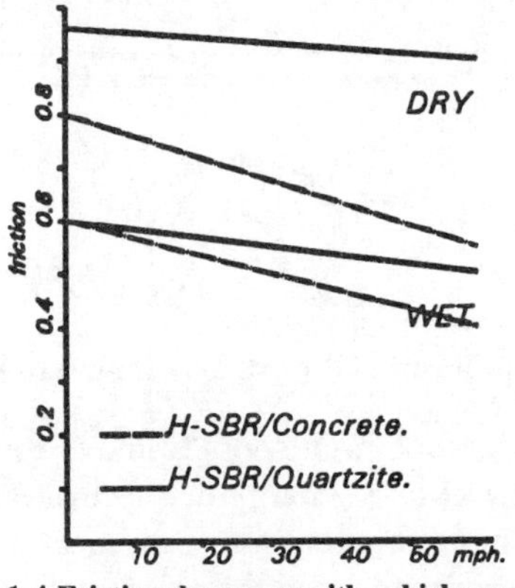

Fig.1.4 Friction decreases with vehicle speed

Quartzite is a rough, harsh surface adequately described by surface A; while polished concrete is a poor surface classified as between C and D. Tests on tyres show that the effective friction number is higher for a lightly loaded tyre than for one operating at greater normal force.

Hydroplaning may occur when a wheel rolls on a surface with standing water.

Effective movement of the wheel requires that a film of water is removed from the contact area before the intimate contact necessary for the development of longitudinal or lateral forces can take place. There is a limit to the amount of water that the tread can pump away and once this is exceeded a wedge of fluid builds up to separate the tyre from the road.

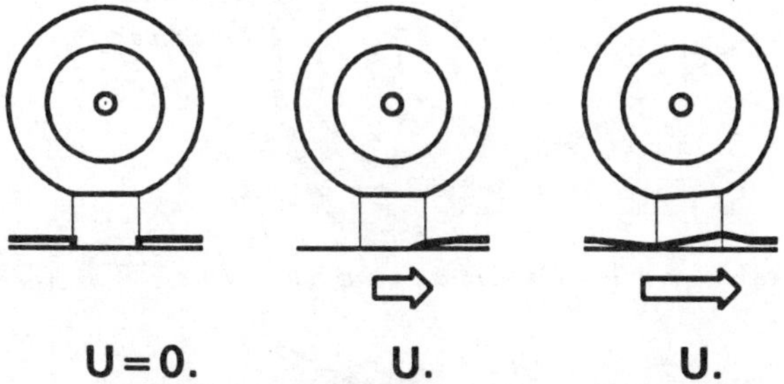

Fig.1.5 A wedge of water builds up between the tyre and the road as vehicle speed increases

This mechanism is a well known hydro-mechanics phenomenon and when it is fully established the wheel may spin down and be unable to regain a rotating state. During this time no control forces are available. Several factors contribute to hydroplaning. For a smooth tyre it can be shown that the speed above which it may start is proportional to $\sqrt{p}$. Tread width is also a factor since the lateral shear forces in the contact patch tend to force water toward the centre of the patch and induce hydroplaning. A pattern of widely spaced tread blocks with a large central channel will help to disperse the free water.

1.3 THE TYRE AXIS SYSTEM

A right handed axis set with the origin located at the centre of tyre to road contact, the x axis pointing forward and the z axis perpendicular to the road and positive in an upward direction is frequently employed in vehicle dynamics. A sketch is given in Fig. 1.6. Note that the spin axis of the wheel is perpendicular to the wheel disk at all times.

Steer is a rotation of the wheel disk around the z axis and is positive for a clockwise rotation around the z axis.

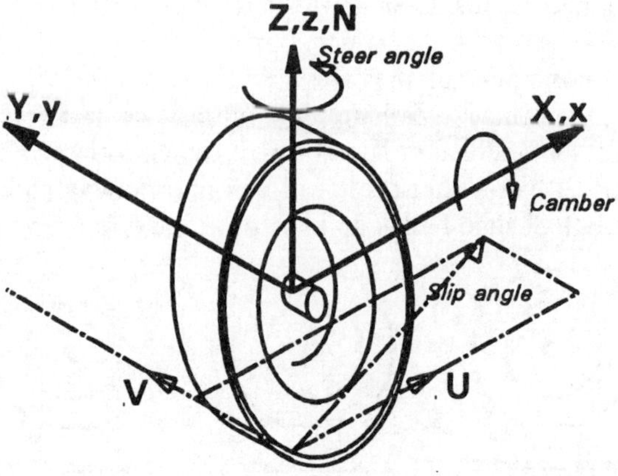

Fig.1.6 Tyre axes may be defined at the wheel hub or the centre of ground contact

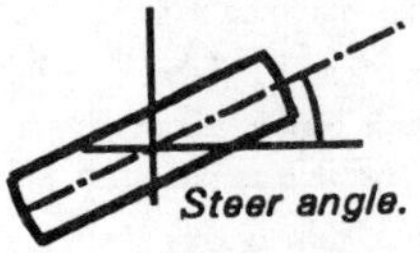

Slip is a vector formed by the lateral and longitudinal velocities of the system. Positive velocities at the contact patch show the tyre moving in a direction equivalent to a negative steer angle.

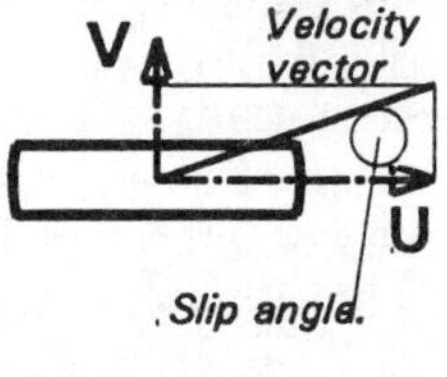

Attitude angle. In any vehicle control situation a slip angle is generated and steering may be present. The attitude angle is used to compute lateral force and aligning moment of the tyre.

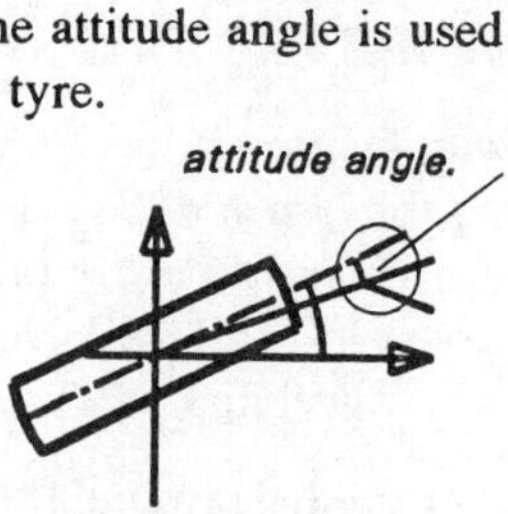

Attitude angle = Steer angle - Slip angle

Camber is a rotation around the *x* axis.

1.4 EQUIVALENCE OF WHEEL CENTRE AND GROUND FORCES

Tyre test data are usually obtained by measuring the forces and moments required to restrain the wheel hub under a given test condition. That data is directly applicable to many problems; however, when the forces acting within a steering linkage are required then it is convenient to replace the measured axle restraints by an equivalent set of forces and moments with the steering axis as reference. Figure 1.7 shows the two systems of forces and moments.

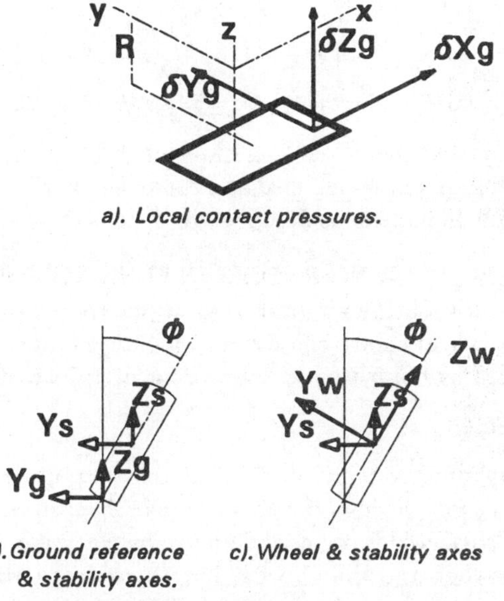

Fig.1.7 Relations between wheel and vehicle axes, assuming a rigid wheel disk

Let X_g, Y_g, Z_g be the local ground contact pressures. Then from Fig.1.7(a) the relations between the local pressures and the total forces and moments are obtained by integration over the contact area.

$$X_g = \Sigma dX_g \qquad L_g = \Sigma y dZ_g$$

$$Y_g = \Sigma dY_g \qquad M_g = \Sigma - x dZ_g$$

$$Z_g = \Sigma dZ_g \qquad N_g = \Sigma (x dY_g - y dX_g)$$

The ground forces and moments are related to the vehicles axes as shown in Fig.1.7(b).

$$X_s = X_g \qquad N_s = L_g - Y_g R\cos\phi - Z_g R\sin\phi$$
$$Y_s = Y_g \qquad M_s = M_g + X_g R\cos\phi$$
$$Z_s = Z_g \qquad N_s = N_g + X_g R\sin\phi \qquad (1.1)$$

The relations between stability axes and wheel which are required to convert test data measured at the axle to a form suitable for stability studies are obtained from Fig. 1.7(c).

$$X_s = X_w \qquad L_s = L_w \qquad (1.2)$$
$$Y_s = Y_w\cos\phi - Z_w\sin\phi \qquad M_s = M_w\cos\phi - N_w\sin\phi$$
$$Z_s = Z_w\cos\phi + Y_w\sin\phi \qquad N_s = N_w\cos\phi + M_w\sin\phi$$

Determination of the forces and moments around the steering axis or kingpin, which is generally inclined at a compound angle is best undertaken by the transformation methods described in Chapter 2.

1.5 ROLLING RADIUS IN FREE ROLLING

The relation between the angular velocity of the wheel and the linear velocity of the wheel centre can define a rolling radius for the tyre. This radius is affected by normal force, tractive force, and lateral force.

$$R_e = U/\Omega \qquad (1.3)$$

As the wheel is rolled forward on the road each part of the circumference is flattened as it passes through the contact area. Shear stresses are generated at the contact surface due to the difference between the arc length of the free circumference and the chordal length represented by the contact surface.

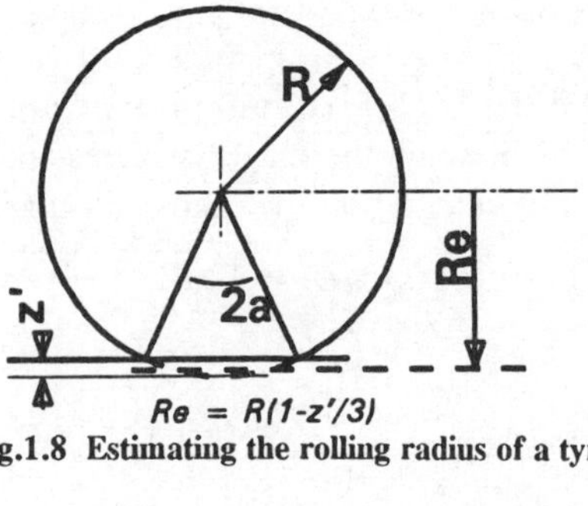

Fig.1.8 Estimating the rolling radius of a tyre

An initial estimate of rolling radius can be made. In this analysis the part of the tyre perimeter subtending the arc of contact is replaced by a straight line of similar length. The rolling radius is the distance of this line from the wheel centre given the same angle of contact.
From the figure the tyre deflection is

$$z' = R\{1-\cos(a)\} \tag{1.4}$$

The arc length of the undeflected tyre in the contact region is

$$2\underline{l} = 2Ra \tag{1.5}$$

The effective rolling radius, R_e, is obtained by calculating length *A-A*, which equals this arc.

$$R_e = Ra/\tan(a) \tag{1.6}$$

Substituting for angle a from equation (1.4)

$$R_e = R - z'/3 \tag{1.7}$$

1.6 ROLLING RADIUS WITH LONGITUDINAL FORCES

Braking and driving forces are resisted by additional shear forces along the contact length. As a result of these forces the rolling radius changes. Ultimately the wheel will either spin without forward motion, or slide without rotation; for small changes in longitudinal force the change in rolling radius is linear, the typical changes are shown in Fig. 1.9.

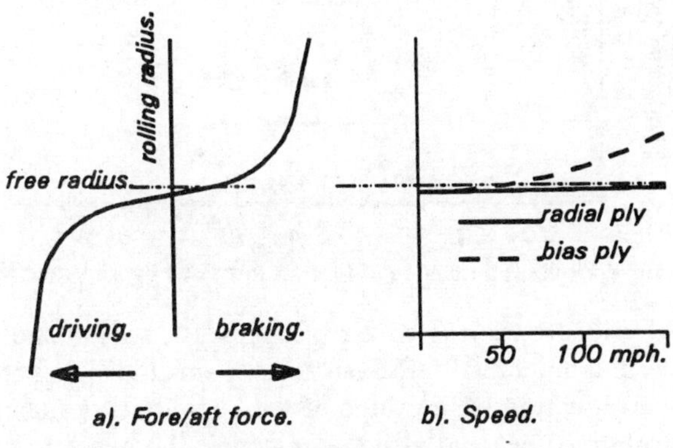

Fig.1.9 Rolling radius vs fore/aft force and speed

1.7 THE EFFECT OF SPEED ON ROLLING RADIUS

As the rotational speed of the wheel increases the tread band tension rises due to centripetal acceleration. The resulting growth of the equatorial band is more noticable in bias tyres than with radial tyres where the tread band is relatively inextensible. Continual movement of the tread through the contact region provides an excitation to the tread area and as the road speed increases a resonance condition may appear in which the deformation will travel around the tyre at the speed of rotation giving rise to a standing wave. At this time there is a massive increase in the power consumption of the tyre leading to failure if the condition is maintained. The standing wave phenomenum occurs mainly in bias ply tyres.

1.8 TYRE SLIP

The previous sections of the text describe the rolling radius of a tyre and the way this radius is affected by speed and longitudinal force. These changes in rolling radius are the external signs of slip within the contact region of the tyre. Slip occurs at all times during the rolling process. Slip may be defined as a vector.

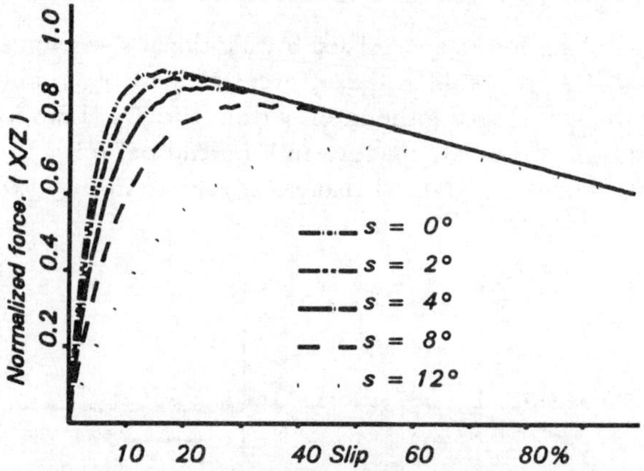

Fig.1.10 Tyre slip is a function of fore/aft force and the slip/steer angle

A typical slip diagram is given in Fig.1.10, which shows a linear relation between longitudinal force and slip for small levels of applied force followed by a reduction in the slope of the curve until a maximum force occurs at around 20 percent slip, after which the available force falls rapidly to a lower equilibrium level.

Slip = *(U-v)/U* (1.8)

For the case of rolling in the longitudinal direction, then

Slip = *1-v/U* (1.9)

When the wheel rolls at an angle, *s*, to the plane of the wheel disk then the slip becomes

Slip = *sin(s)* (1.10)

The concept of slip, or relative motion within the contact patch is applicable to both the steered and rolling conditions. The mechanism by which the tyre reacts against the road is similar for steering, longitudinal forces and combinations of those conditions.

1.9 THE STEERED, ROLLING TYRE

Rolling the wheel forward in contact with the road causes the forward portion of the contact area to be continuously renewed while the rear is released. The continual entry of the tread at the front of the contact patch requires that the deflection and slope of the tread band are similar immediately before and after the start of contact.

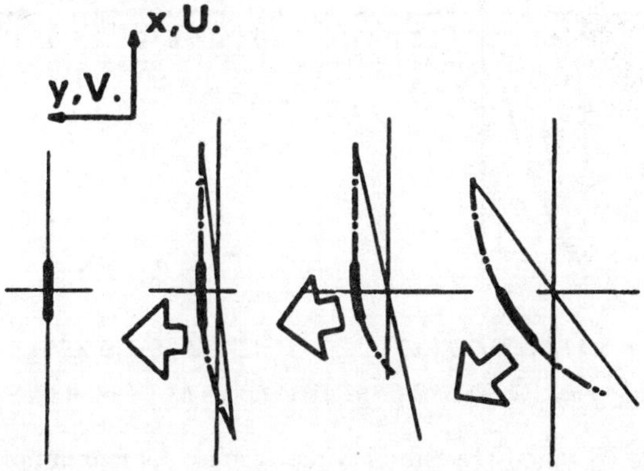

Fig.1.11 Tyre lateral deflection caused by steer or lateral slip gives rise to F&M

Figure 1.11 shows some sketches of the centreline of contact for a model tyre. Starting on the left the first diagram shows the centreline of the contact patch for the freely rolling tyre. The next sketch illustrates the lateral deflection due to a small steer angle - note the continuity of slope at the front of the patch and the offset of the lateral force due to the assymetric deflection of the tread. This is followed by two diagrams for increasing values of steer angle. The leading edge of the contact patch moves farther from the wheel disk in order to maintain continuity here, while the deflection at the rear is limited by friction. The *Y* vector indicates the change in magnitude of the lateral force while the product Y^*x gives an indication of the aligning moment.

For small steer angles the lateral force vs steer angle relation is linear but as the angle is increased the response falls off due to sliding at the rear of the contact patch where the deflection of the tread is greatest. The moment around the *z* axis, the aligning torque, is the moment of the forces in the ground plane around the steering axis.

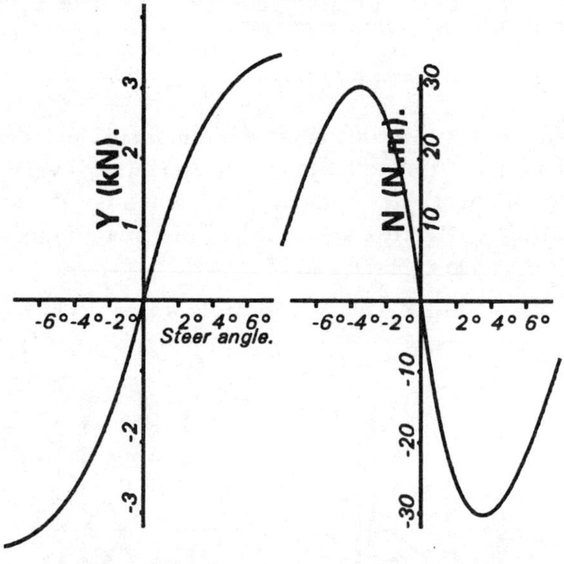

Fig.1.12 Typical F&M cuves for a tyre on a test rig

The angle at which the lateral force reaches its maximum will vary, for a racing car rear tyre lateral force peaks at about 6 degrees, while for passenger vehicles the maximum force may occur about 18 degrees. In vehicle control analysis it is not the lateral force itself which is important, but the rate of change of force with angle, dY/ds, the lateral force coefficient. When this is high a vehicle will respond rapidly to steering and other disturbances, but the response is slower and of smaller amplitude for low values of *dY/ds*. Thus when the vehicle is already in a turning situation the responses around that condition, which are dependent upon the local values of *dY/ds*, will be slower than from the zero trim state because of the reduction in lateral force coefficients. This will be discussed in detail later.

It is convenient to present tyre test data polynomial form. Note that the use of the absolute value of attitude angle in the 2nd order term ensures symmetry in the first and third quadrants

$$Y = C_1 s + C_2 s\,|\,s\,| + C_3 s^3 \qquad (1.11)$$

Each of the coefficients is a function of the normal force, Z. A curve fitting routine is described later.

1.10 THE CAMBERED WHEEL

Camber is a rotation around the *x* axis. Car and truck tyres are designed to operate in a nominally upright position and will tolerate only small amounts of camber. Motor cycle tyres run at large angles of camber during cornering because the single track vehicle uses the roll mode for stability and must always be in equilibrium around that axis.

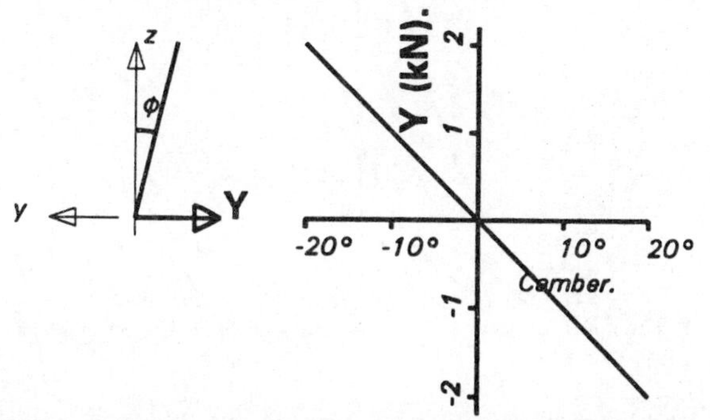

Fig.1.13 Lateral force is generated when a tyre runs with a camber angle

If the cambered wheel is considered as a thin disk in contact with the road, the initial contact line is part of an ellipse. As the wheel is rolled forward the contact becomes a straight line parallel but offset from the direction of rolling. Thus, cambering the disk will produce a lateral force with no moment. However, the width of the tyre causes a moment to arise due to the non-uniform deformation of the contact patch.

Camber sensitivity for a car tyre is much less than the steer sensitivity - a typical value for a radial tyre will be around 1/20th of the steer value.

1.11 CAMBER AND STEERING

Most suspensions impart some degree of camber to the road wheel as the vehicle body rolls in a turn; thus, the wheels operate under a combination of steering, slip, and camber. Tests indicate that the effect of camber is to shift the origin of the lateral force curve without changing either the slope or the maximum/minimum values. From the Fig.1.14 it is clear that the previous formula for lateral force can still be employed, provided that an allowance is made for camber.

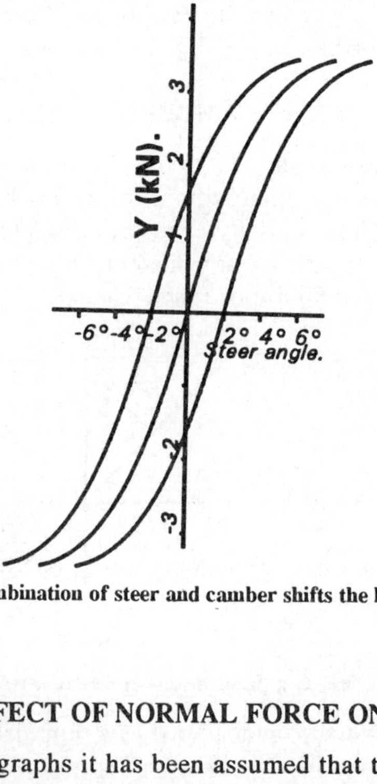

Fig.1.14 The combination of steer and camber shifts the lateral force curve

1.12 THE EFFECT OF NORMAL FORCE ON STEERING

In the previous paragraphs it has been assumed that the normal force, or axle load, has remained fixed during the test. During vehicle operations the normal force changes, for example, due to roll or braking, and this redistribution of loading causes changes in the force and moment generated by steering. This interaction between the normal and lateral responses is frequently used to modify the handling responses. A plot of the effect is shown in Fig.1.15.

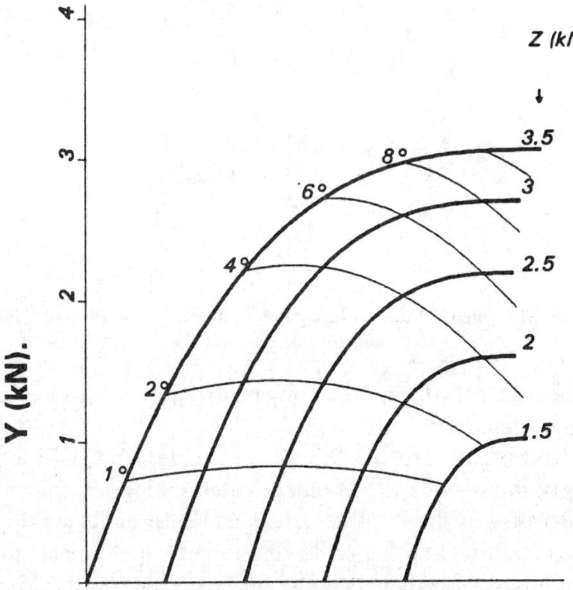

Fig.1.15 A carpet plot showing the effect of normal force and steer angle on lateral force

1.13 THE EFFECT OF LONGITUDINAL FORCE ON LATERAL FORCE

Braking or driving forces change the lateral force generated at any attitude angle.

The change occurs because the longitudinal force moves the contact patch, mainly in the direction of the braking or tractive force but the lateral displacement of the contact length relative to the wheel plane will experience some change. A small alteration in length of the contact patch also occurs due to the presence of the additional strain in the tread.

Figure 1.16 shows the effect in an approximate manner. These movements may be reconciled with the longitudinal and lateral stiffnesses of the tyre.

1.14 AN APPROXIMATION FOR LATERAL AND LONGITUDINAL FORCE INTERACTION

An estimate of the lateral force available at a specified attitude angle is possible if it is assumed that the tyre may reach a limiting force condition

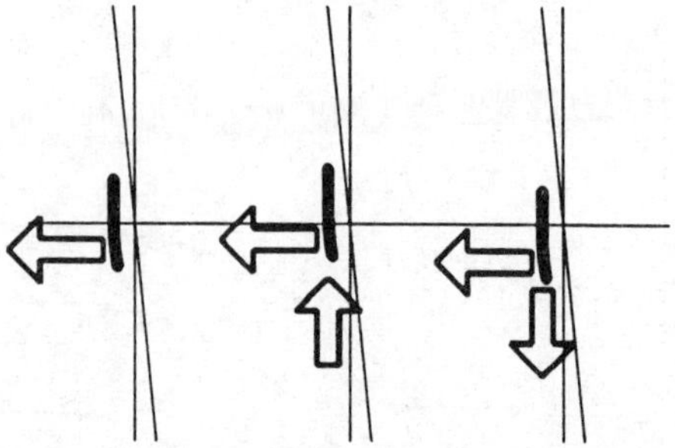

Fig.1.16 Movement of the contact patch under a combination of lateral and fore/aft forces

under a combination of applied longitudinal force and lateral force generated kinematically.

Because braking or driving the wheel is a demand, while the lateral force is simply the result of the contact patch taking up a position offset from the wheel disk as the local velocities and steer angle are applied, then in an extreme condition braking or driving forces will dominate the situation and no lateral force can be generated by the tyre whatever its slip angle.

In an intermediate condition the lateral force and moment will be reduced in the presence of braking or tractive forces. The left side of Fig.1.17 is a steer angle vs lateral force curve for a freely rolling tyre. The construction to modify lateral force is as follows: the lateral force at the selected attitude angle (1 kN at 1.4 degrees) is projected horizontally onto the vertical axis in the centre of the diagram. The horizontal axis of the right hand side represents longitudinal force. Following the friction ellipse concept an ellipse is constructed with this value of *Y* and the maximum value of fore/aft force as the axes. The lateral force at a fixed attitude angle in the presence of driving or braking forces is obtained by reading the local ordinate.

The ellipse describing the change in lateral force at a fixed steer angle is given in equation (1.12)

$$Y_z = Y\sqrt{\{1 - (X/X_m)^2\}} \quad (1.12)$$

1.15 PNEUMATIC TRAIL

In many cases it is convenient to express the aligning moment as the product of the lateral force and some displacement from the tyre centre. The term *pneumatic trail* is used for this lever arm. When the initial slopes

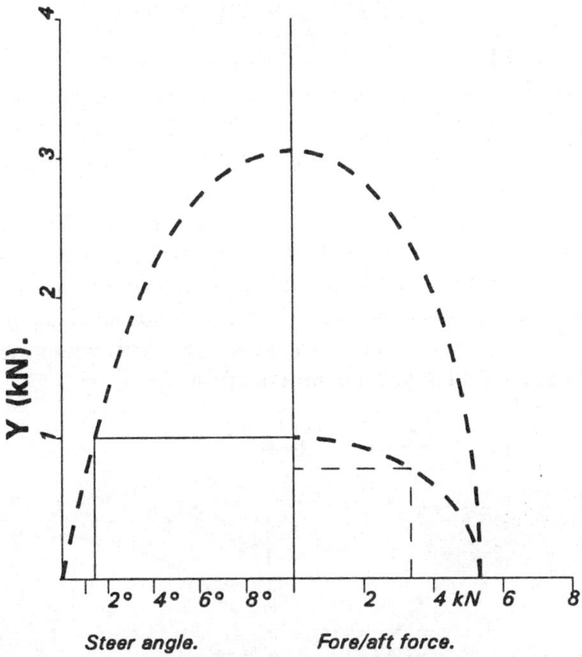

Fig.1.17 The friction ellipse shows the available lateral force with fore/aft force present

of the lateral force and moment curves are known then the trail for zero steer angle is

$$x = (dN/ds)/(dY/ds) \qquad (1.13)$$

At larger angles a simple expression for the pneumatic trail is

$$x = (dN/ds)/(dY/ds)(1 - |s/s_{max}|) \qquad (1.14)$$

When the tyre is operating under a combination of lateral and longitudinal forces the contact patch is shifted both longitudinally and laterally (Fig.1.16). The moment around the *z* axis of the wheel disk is a function of both fore/aft and lateral forces. Thus, for combined braking and steering the centre of pressure will shift rearwards relative to the wheel centre, providing an increase in the negative component due to lateral force ($Y*x$) while the component from braking ($X*y$) is positive; a shift from a restoring, or negative, moment to a positive moment occurs as braking is applied. As a tractive force is applied to the steered wheel then the restoring moment (-ve) is increased.

1.16 TYRE IRREGULARITY

A tyre does not have the clearly defined mechanical properties of a mechanical test specimen and it is to be expected that many of the characteristics will vary as the tyre rotates. Typically a tyre will show variations in lateral force, aligning moment, radial force, and stiffness during a cycle of rotation.

By measuring lateral force when the tyre is rolled in positive and negative directions the effects of ply offset and conicity can be demonstrated; some typical results are shown in Fig.1.18. These characteristics, when considered for the four tyres on a vehicle, can cause steering pull, apparent offset of the steering wheel when the vehicle is running in a straight line, and a crab-like effect.

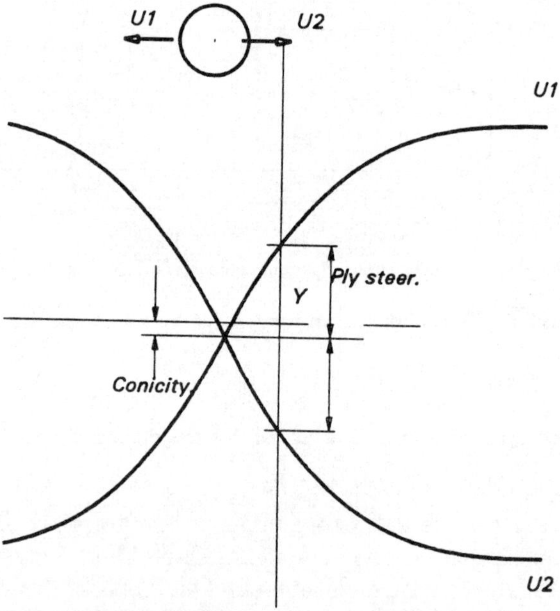

Fig.1.18 Conicity and ply steer are determined by superimposing results from tests with opposing directions of rotation

A set of 12 similar tyres tested for static lateral stiffness and for lateral cornering force characteristic gave the results shown in Table 1.1. The larger standard deviation obtained for the static tests appears to represent the less sophisticated test rig used for those tests. These results give a guide to the standards of tyre construction available.

Table 1.1 Mean values and standard deviations of cornering stiffness and static lateral stiffness at various normal forces (P215/75 R 14 tyre; sample size 15)

Z		*dY/ds* (*lbs/deg*)		*dY/dy* (*lbs/in*)	
(lb)	*(lb/in²)*	*mean*	*s.d.*	*mean*	*s.d.*
	20	197	7.5	462	33
860	26	219	7.0	471	34
	32	207	5.7	500	38
	20	180	8.3	417	56
1433	26	232	7.8	480	37
	32	242	5.7	530	50
	20	140	7.8	420	42
2006	26	197	6.3	489	41
	32	226	8.0	548	47

1.17 INFLATION PRESSURE

The tyre relies upon internal pressure for its structural stability and retention on the wheel. Greater inflation pressure increases the mechanical stiffnesses while reducing the area of contact for a given normal force. Thus, a small rise in pressure from the standard value will produce a higher steering force characteristic, while the aligning moment will not change or may even be reduced. Over-inflation leads to rapid wear at the centre of the tread, while under-inflation is a major reason for structural failure due to overheating.

1.18 ROLLING LOSSES

As a tyre rolls forward the mechanical energy from hysteresis is converted into heat. Although energy is usually expressed in Joules it has become conventional in tyre studies to represent the energy as the work done by a fictitious force acting at the rolling radius of the tyre/road interface. The energy absorbed by the tyre in any condition is the difference between work in and work out.

$$\begin{aligned} dH/dt &= \text{Power in} - \text{Power out} \\ &= T - XU \end{aligned}$$

Hence

$$E = T/U - X \tag{1.15}$$

When a torque is transmitted by a tyre the rolling loss changes in a manner reminiscent of the effect of torque on rolling radius as seen in Fig.1.28, thus confirming the expression given in equation (1.15). Note that the minimum energy loss occurs when the wheel is driven by a light torque, in the example shown the minimum loss corresponds to an applied torque of 150 Nm.

The effect of steering a tyre is to generate a side force and a lateral displacement. Energy is dissipated by the movement of the tyre on the road. The lateral force vs deflection characteristic of the tyre is linear for small displacements and, hence, the work done in deflecting the tyre can be estimated.

$$E = Ky^2/2 \tag{1.16}$$

The force developed due to lateral deflection can be related to the force developed by steering and thus the energy dissipated during steering can be expressed in terms of familiar tyre characteristics.

$$Y = Cs = Ky \tag{1.17}$$

Thus in the linear region of tyre performance the energy lost can be expressed in terms of the attitude angle.

$$E = (C^2/K)s^2/2 \tag{1.18}$$

1.19 ANALYSIS OF TEST DATA

Tyre test data are usually required for use in handling simulations and for this purpose it is conveniently presented in polynomial form.

$$Y = C_0 + C_1 s + C_2 s\,|\,s\,| + C_3 s^3 \tag{1.19}$$

This equation is chosen because apart from the zero order term it is symmetrical in the first and third quadrants. Note that this data is collected at nominally constant values of axle load. When the test is repeated at a number of loads each of the coefficients of equation (1.19.) may be expressed as a quadratic function of normal force, thus for n=0 to 3

$$C = A_0 + A_1 Z + A_2 Z^2 \tag{1.20}$$

The routine for obtaining a lateral force equation now consists of two curve fitting exercises. Coefficients of equation (1.19) are obtained from data taken at various constant normal forces, and these coefficients are then assembled into sets to give further constants described by equation (1.20).

1.20 AN OUTLINE OF THE LEAST SQUARES METHOD

The form of the lateral force equation given in equation (1.19) is not immediately suitable for curve fitting, so it is convenient to make all data sets positive and then use the cubic polynomial

$$Y = C_0 + C_1 s + C_2 s^2 + C_3 s^3 \tag{1.21}$$

As a result of testing the tyre at a constant normal force, M sets of data (Y,s) are available. Assume that an equation such as given above (equation (1.21)) is to be used to represent the lateral force. At each measured value of steer angle, s, the difference between the measured and calculated lateral force is obtained, the difference is then squared and summed over the whole set. Equation (1.22) is the result of this operation.

$$S = \{Y(\text{observed}) - Y(\text{equation})\}^2 \tag{1.22}$$

The equation (1.21) now replaces Y(equation)

$$S = (Y_i - (C_0 + C_1 s + C_2 s^2 + C_3 s^3))^2 \tag{1.23}$$

This function is minimized by taking the partial differential of the function with respect to each of the coefficients C_0 to C_3 and equating the resulting expressions to zero. This operation gives a set of simultaneous equations which may be solved to give the optimum values for the coefficients.

$$\begin{aligned} \partial T/\partial C_0 &= -2\sqrt{S} = 0 \\ \partial S/\partial C_1 &= -2s_i\sqrt{S} = 0 \\ \partial S/\partial C_2 &= -2s_i^2\sqrt{S} = 0 \\ \partial S/\Omega C_3 &= -2s_i^3\sqrt{S} = 0 \end{aligned} \tag{1.24}$$

These equations give the values of the coefficients. When rewritten in matrix form their suitability for computer solution is apparent.

$$\begin{bmatrix} M & \Sigma s_i & \Sigma s_i^2 & \Sigma s_i^3 \\ \Sigma s_i & \Sigma s_i^2 & \Sigma s_i^3 & \Sigma s_i^4 \\ \Sigma s_i^2 & \Sigma s_i^3 & \Sigma s_i^4 & \Sigma s_i^5 \\ \Sigma s_i^3 & \Sigma s_i^4 & \Sigma s_i^5 & \Sigma s_i^6 \end{bmatrix} \begin{bmatrix} C_0 \\ C_1 \\ C_2 \\ C_3 \end{bmatrix} = \begin{bmatrix} \Sigma Y_i \\ \Sigma Y_i s_i \\ \Sigma Y_i s_i^2 \\ \Sigma Y_i s_i^3 \end{bmatrix} \tag{1.25}$$

This process is repeated for each of the values of Z for which data is recorded with the result that N sets of values of (C_0,C_1,C_2,C_3,Z) are available for a second 'least squares' equation

$$S = \{Y - (A_0 + A_1 Z + A_2 Z^2)\}^2 \tag{1.26}$$

Optimum values of the coefficients are calculated from the simultaneous equations (1.26), which are given in matrix form

$$\begin{bmatrix} N & \Sigma Z_i & \Sigma Z_i^2 \\ \Sigma Z_i & \Sigma Z_i^2 & \Sigma Z_i^3 \\ \Sigma Z_i^2 & \Sigma Z_i^3 & \Sigma Z_i^4 \end{bmatrix} \begin{bmatrix} A_0 \\ A_1 \\ A_2 \end{bmatrix} = \begin{bmatrix} \Sigma Y_i \\ \Sigma Y_i s_i \\ \Sigma Y_i s_i^2 \end{bmatrix} \tag{1.27}$$

1.21 SINE FUNCTION TYRE MODELS

A recent development in the fitting of test data is the use of sine functions based on the NAG library of mathematical methods available on many digital computers. Measured data is represented by thirteen (13) coefficients which form stiffness (B), shape (C), peak (D), and curvature (E) factors. The normal force, Z, is the independent variable used in constructing polynomials for the various factors.

Lateral force formula

$$Y = D\sin\{C\arctan(B\Phi)\} + \delta S_v$$

where

$$\Phi = (1-E)(\alpha+\delta S_h) + (E/B)\arctan\{B(\alpha+\delta S_h)\}$$

$$B = \{a_3 \sin(a_4 \arctan(a_5 Z)\}(1-a_{12} \mid \phi \mid)/CD$$

$$C = 1.3$$

$$D = a_1 Z^2 + a_2 Z$$

$$E = a_6 Z^2 + a_7 Z + a_8$$

$$\delta S_v = a_9 \phi$$

$$\delta Sh = (a_{10} Z^2 + a_{11} Z)\phi$$

Aligning moment formula

$$N = D\sin\{C_1 \arctan(B_1 \Phi)\} + \delta S_v$$

$$C_1 = 2.40$$

$$\Phi = (1-E_1)(\alpha+\delta S_h) + (E_1/B_1)\arctan\{B_1(\alpha+\delta S_h)\}$$

$$B_1 = (a_3 Z^2 + a_4 Z)(1 - a_{12} \mid \phi \mid)/(CD.e^{a5Z})$$

$$E_1 = E/(1 - a_{13} \mid \phi \mid)$$

Brake force formula

$$X = D\sin\{C_2 \arctan(B_2 \Phi_2)\}$$

$$\Gamma = fore/aft\ slip.$$

$$\Phi_2 = (1 - E)\Gamma + (E/B_2)\arctan(B_2\Gamma)$$

$$B_2 = (a_3 Z^2 + a_4 Z)/(CD.e^{a5Z})$$

$$C_2 = 1.65.$$

The successful use of these NAG routines requires close approximations to the final values if the routine is to be convergent.

1.22 A GENERIC TYRE MODEL

Tyre data is frequently difficult to obtain and when available only covers a small part of the vehicle operating conditions. It is often necessary to make inspired guesses.

The flow chart (Fig.1.19) shows the steps to be taken when constructing a subroutine for F&M. The following calculations and decisions must be taken for each tyre at each step during a simulation.

- Is the tyre in contact with the ground ?
- Friction decreases with speed.

- The maximum shear force limits the available traction/braking.
- Peak attitude angle is a function of friction and normal force.
- Is the attitude angle > peak value?

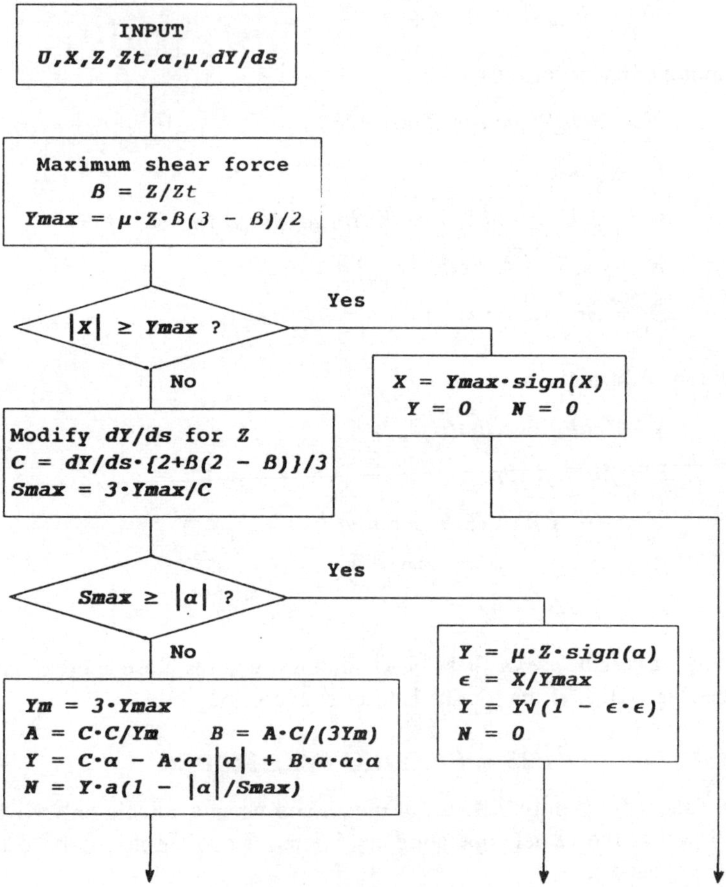

Fig.1.19 Flow chart

A typical coding would be:

```
    DELAY = DELTA*U*Ky/DYDS
C       relaxation = UK/C
C       DELTA = time step
    FZT = FZ + DZ
    IF FZT<=0. THEN 200
C       Is the tyre in contact ?
    FRIC = MU*(1-.01*U)
C       friction falls with speed
    ZN = FZT/ZTEST
C       F&M are load dependent
    YMAX = FRIC*FZT*ZN*(3-ZN)/2
    IF FX = 0 THEN 100
C       skip if tyre free rolling
    XMAX = FRIC*FZT
    IF ABS(FX)>=XMAX THEN 300
C       no F&M if wheel locked/spinning
    YMAX=YMAX*SQRT(1-(FX/XMAX)^2)
C       modify Ymax for drive/brake
100 CONTINUE
C       start F&M calculations
    YM = YMAX/3.
    C = DYDS*(2.+ZN*(2.-ZN))/3.
C       initial slope is load dependent
    SMAX = YM/C
C       attitude angle for Ymax
    SMOD = ABS(SO)
    IF SMOD>SMAX THEN 110
C       is actual attitude angle beyond range of formula ?
    FY = C*SO*(1-C*(SMOD-C*SO^2/(3*YM))/YM)
C       generic cubic for Y
    FY = FY-(FY-FYlast)*EXP(-DELAY)
C       Y reduced by relaxation length
    TRAIL = DNDS/DYDS*(1-SMOD/SMAX)
C       pneumatic trail
    SAT = FY*TRAIL
    RETURN
110 FY = YMAX*SGN(SO)*SQRT(1-(FX/XMAX)^2)
    SAT = 0.
    FX = FX
    RETURN

300     FX=XMAX*SGN(FX)
    FY = 0
    SAT = 0
    RETURN
```

1.23 TRANSIENT TYRE CHARACTERISTICS

When the tyre is stimulated by a non-steady input such as steering or variation in normal force the lateral force and moment change from those of the steady state. The application of a step steer input to a tyre is shown in Fig.1.20, where the gradual build up of force with distance is shown. The subtangent *a-c* is the relaxation length. This distance, or time, dependency is a function of the mechanical characteristics of the tyre.

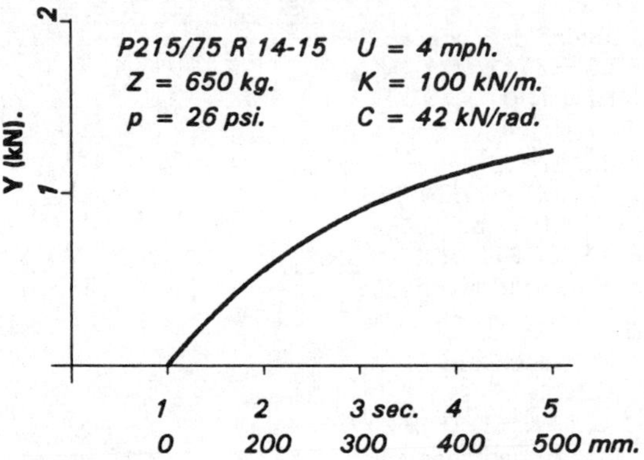

Fig.1.20 The lateral force 'relaxation length'

Some experimental results shown in Fig.1.21 give the phase and amplitude of force and moment when a tyre is steered sinusoidally. The abscissa for these diagrams is the ground wavelength of the oscillation.

Figure 1.22 shows that the result of applying a sinusoidal normal displacement to a tyre steered at a constant angle is that a non sinusoidal lateral force is generated. The applied displacement is so large that the tyre leaves the road during the cycle. There is no satisfactory model which can be used for the study of the lateral response of the steered tyre to variation in normal force at high frequency. At low frequencies such as occur in vehicle handling the assumption of quasi-steady state response appears reasonable and then the effect of changing load may be simulated in a manner similar to that described below.

1.24 A LATERAL SPRING MODEL FOR TRANSIENT RESPONSES

Tyre lateral stiffness is described in section 1.2, and the development of lateral force due to attitude angle in section 1.8. These events do not occur

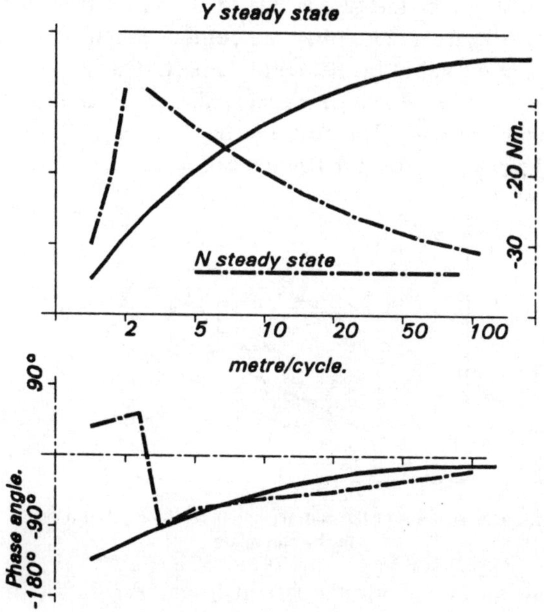

Fig.1.21 F&M responses to sine steering

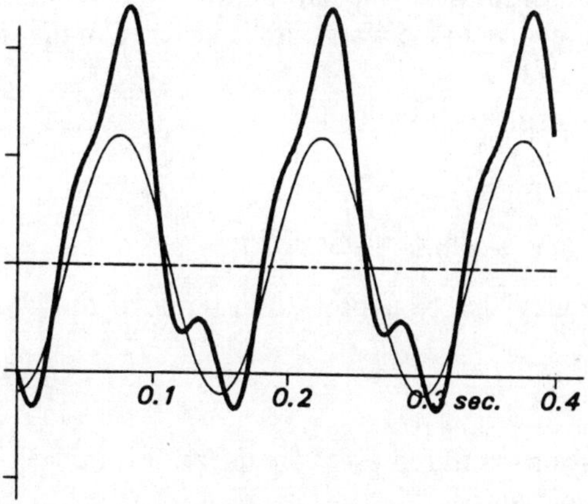

Fig.1.22 This graph shows the result of a sine variation in axle height

in isolation and, clearly, lateral deflection occurs between wheel and contact area when the tread contact patch will move relative to the wheel rim and the velocity of this displacement will contribute to the tyre slip angle.

The attitude angle of the tyre is due to a combination of steering and sideslip angle. Figure 1.23 shows the contact patch of a tyre which is steered about a point not at the centre of contact. This is a typical situation for the front wheel of a vehicle where this point, p, is the intersection of the steering axis with the road. The centre of the contact patch will be taken as representing the area. A lateral deflection, y', is assumed.

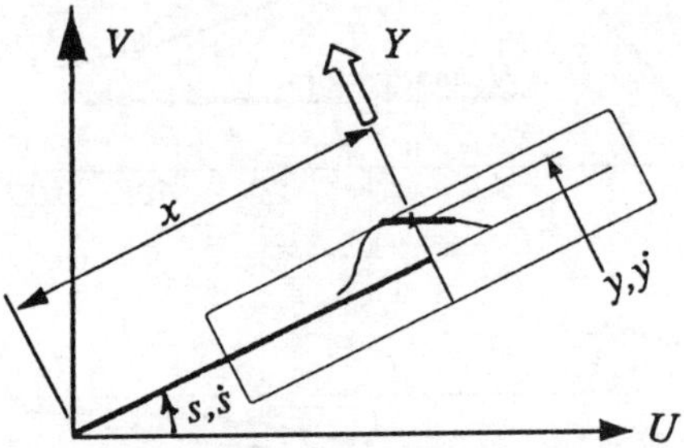

Fig.1.23 Lateral velocity of the contact patch relative to the wheel is a term in the slip angle

For a small attitude angle the lateral force, $Y(s)$ is the product of the lateral force characteristic, dY/ds, and the attitude angle which has been defined as the sum of the steer and slip angles. Note that lateral velocity of the contact patch relative to the wheel rim is included in the slip angle term. Let

$$C = dY/ds \mid_{s=0} \tag{1.27}$$

Then

$$Y = C(s -(V + xds/dt + dy'/dt)/U \tag{1.28}$$

Lateral force may also be expressed in terms of the lateral deflection and stiffness

$$Y = Ky' \tag{1.29}$$

These expressions both represent the lateral force developed by a tyre, thus they are always equal

$$Ky' = C(s - (V + xds/dt + dy'/dt)/U \tag{1.30}$$

Let

$$\sigma = UK/C$$

Then

$$\sigma y' + y' = Us - xs - V \qquad (1.31)$$

This model is now used to demonstrate the 'relaxation length' resulting from a step steering input.

The equation to be solved is derived from equation 1.31 by considering thatthe steering takes place around the centre of the wheel.

The initial conditions are

$$V = 0;\ y' = 0;\ s = S \quad \text{when } t = 0$$

Hence

$$dy'/dt = Us - \sigma y' \qquad (1.32)$$

Assume

$$y' = A(1 - e^{-\sigma t})$$

Then

$$dy'/dt = \sigma A e^{-\sigma t}$$

Substitute for y' and dy'/dt, together with the values of y' and s when $t = 0$.

$$y' = (sC/K)(1 - e^{-\sigma t}) \qquad (1.33)$$

The lateral force is then the product of the deflection and the lateral tyre stiffness

$$Y(t) = sC(1 - e^{-\sigma t}) \qquad (1.34a)$$

Then

$$Y(t) = Y_{ss}(1 - e^{-\sigma t}) \qquad (1.34b)$$

Lateral force may be written in terms of distance moved, since $x = Ut$

$$Y(x) = sC(1 - e^{(-Kx/C)}) \qquad (1.35)$$

The 'relaxation length' is the sub-tangent to this curve. Hence

$$l_r = C/K \qquad (1.36)$$

This model can be used in many simulations where tyre elasticity is required, that is usually when the steering mechanism is part of the system.

Its major limitation is that a point contact is assumed and thus it will not represent short wavelength (high frequency) motion. A ground wave length greater than three times the contact length provides good results. On the other hand the parameters are clearly defined and may be measured quite simply. The extra degrees of freedom needed in this simulation are an additional computing load that is not always welcome. When it is not necessary to study the vehicle at standstill it is possible to modify the steady state lateral tyre forces simply by introducing a delay time of the type shown in equation (1.34b) and also in section 1.21.

1.25 SHIMMY

The term 'shimmy' is frequently used to describe two different phenomena. There is a tendency for a wheel to move on a sinusoidal path, rather than a straight line when it is slowly rolled forward. This is kinematic shimmy. Dynamic shimmy is a vibration around the steering axis which involves the inertia of the wheel and other parts about that axis with the tyre lateral force acting as a damper/exciter as described in section 1.23.

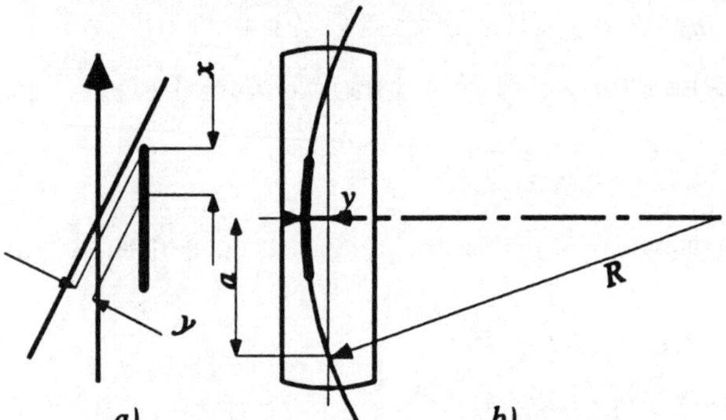

Fig.1.24 Kinematic shimmy occurs at low speed and is due to lateral run out land/or bearing clearance

One of the simpler explanations of kinematic shimmy is now described. A small lateral run out is assumed with the result that the tyre contact path rhas a small angle, s, in Fig.1.24(a). Measuring from the leading edge of contact

$$y = x.sin(s)$$

For small angles

$$s = dy/dx$$

Figure 1.24(b) shows the tyre rolling on a circular arc of radius R, the lateral tyre deflection is y

$$R^2 = a^2 + (R-y)^2$$

or

$$R = a^2/2y$$

As the wheel rolls forward the angle of the tyre centreline changes since the lateral run out varies around the circumference

$$Rds = dx$$

Therefore

$$ds/dx = 2y/a^2$$

It can then be shown that the motion is simple harmonic and that

$$d^2s/dx^2 = -2s/a^2 \qquad (1.37)$$

Dynamic shimmy involves the inertia and stiffnesses of the steering mechanism referred to the steering axis and the damping effect of F&M developed by the laterally flexible tyres. The model is similar to that for the tyre relaxation.

I = polar moment of inertia around steering axis.
K = Torsional stiffness between frame and wheel.
x = mechanical trail, usually negative.
x'= x + pneumatic trail.

$$I\ddot{s} + Ks + Yx + N = 0 \qquad (1.38)$$

Y is defined by equation (1.28), hence

$$I\ddot{s} - Cx.x'\dot{s} + (K + Cx')s = \dot{y}.C/U \qquad (1.39)$$

$$\dot{y}' = Us - x.\dot{s} - \sigma y'$$

When K is zero the system represents a wheel free of springlike restraint around the steering axis. Friction is usually present and this controls the incipient oscillation for small perturbations. However the wheel becomes divergently unstable when the aligning moment coefficient is greater than the moment from lateral force offset from the steering axis; ie when the mechanical trail is positive and opposite in sense to the pneumatic trail.

With a large value of trail the wheel will thrash from side to side in a limit cycle. Bearing clearance allows the tyre to move laterally without control.

Shimmy is always a possibility with a steered wheel, particularly with worn tyres. An early solution to the practical problem was the development of tyres with two widely separated treads and a central groove. Increased friction is also used to stop shimmy.

1.26 SMOOTH ROAD TYRE MODELS

A typical tyre model for use in ride and handling analyses is the familiar spring/mass system with the 'unsprung mass' of the wheel and brake assembly and the tyre normal stiffness as the spring. However there are some problems in assigning a realistic damping factor to this model since the majority of damping seems to occur as scuffing of the contact patch.

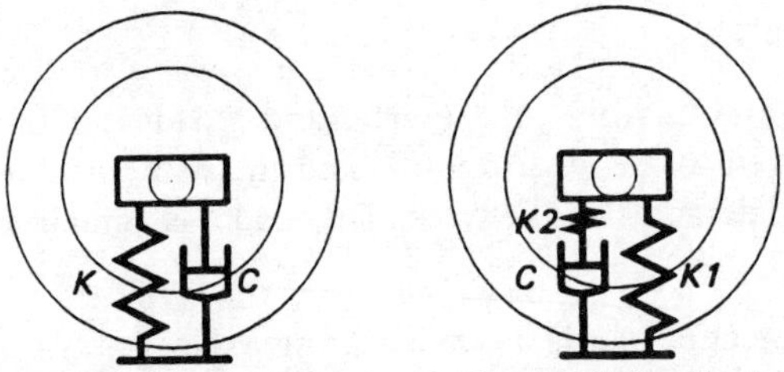

Fig.1.25 Tyre spring models for ride calculations

An alternative is to use the Maxwell or relaxation model of Fig.1.25, in which the damping element is in series with a second spring. This model has the philosophical advantage that the damping is effective at low frequencies but as the input frequency increases the energy absorbed is limited by movement of the secondary spring.

The equation for the spring-mass model is

$$m\ddot{z} + c\dot{z} + kz = f(t) \tag{1.40}$$

For the Maxwell model it is necessary to have a second degree of freedom representing the force across the damper

$$\begin{aligned} m\ddot{z}_1 + k_1\{z_1 - f(t)\} + k_2(z_1 - z_2) &= 0 \\ -k_2(z_1 - z_2) + c\{\dot{z}_2 - f'(t)\} &= 0 \end{aligned} \tag{1.41}$$

1.27 THE TYRE ON ROUGH TERRAIN

One of the disadvantages of the models shown in the previous section is that they do not demonstrate the enveloping qualities of the tyre on short bumps when the vertical response is small and high longitudinal forces are generated.

Use of the single spring point contact model is limited because of the inherent assumption that the terrain upon which the tyre is standing is a plane tangent to the actual surface in the vicinity of the contact point. When the following conditions are met satisfactory results may be obtained.

(a) No step changes in the path.
(b) Least surface wavelength is more than three times contact length.
(c) The plane adequately defines the contact surface.

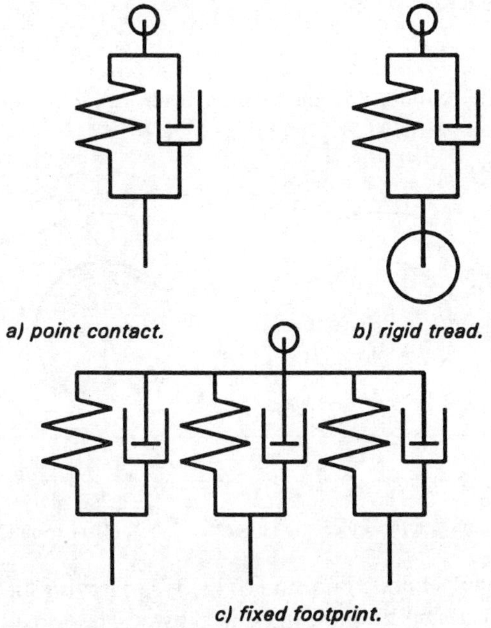

Fig.1.26 Various models for terrain envelopment

A number of modifications of the single spring model have been proposed and these are illustrated in Fig.1.26. The point contact unit reacts to the vertical displacement of the ground and feeds both longitudinal and vertical forces to the wheel hub. The rigid wheel of Fig.1.26(b) modifies the ground profile from that met by the first model. The fixed footprint model may use point contact as shown in Fig.1.26(c) or may include rigid wheels at each spring. The adaptive footprint model of Fig.1.26(d) is a series of radial springs each of which is deflected by the local terrain.

A model uses the single spring model examining a length of ground profile eqivalent to the footprint, the mean height and inclination of the ground are computed and the force vector is perpendicular too this imaginary plane.

A short wavelength obstacle generates both vertical and fore/aft forces at the hub as a tyre passes over it.

In Fig.1.26. the tyre spring is a distributed stiffness constant along the footprint.

Let the total vertical stiffness be K and the distributed spring k. $2l$ is the length of contact

$$Z = k \int h(x).dq \tag{1.42}$$

When the tyre is deflected uniformly, then $h(x)=z$, and the relation between the normal stiffness, K_z, and k is obtained.

$$k = K_z/2l \tag{1.43}$$

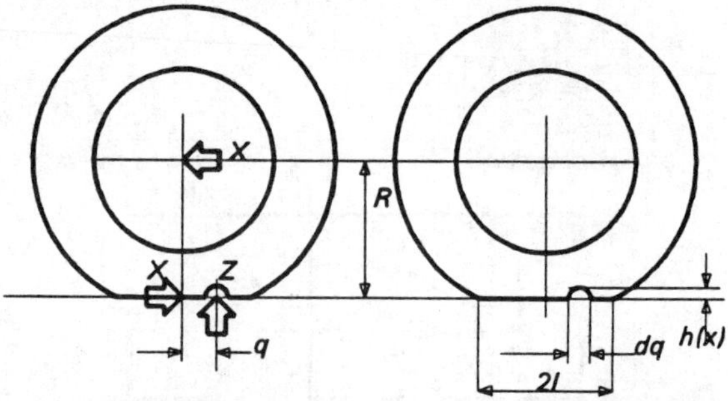

Fig.1.27 A bump causes a local increase in normal force, offset from the wheel centre

In Fig.1.27 the general condition of the tyre carrying an axle load, Z, and passing over a small bump, $h(x)$, is shown. The forces involved are given. The increase in normal force Z is obtained from equation (1.41), this force acts at a distance q from the centreline of contact. The moment produced by the change in normal force is balanced by equal and opposite forces X, separated by R

$$X = Z.q/R \tag{1.44}$$

While a continuous radial spring model can provide a tyre response provided that the terrain can be described deterministically the computational burden is high. An obvious solution is to employ a finite number of radial springs and to define the ground plane at each segment in terms of slope and height. The radial spring stiffness can be estimated from the measured static stiffness and then when an arc of tyre circumference is allocated an actual stiffness is given to each arc. With the ground divided into a number of equivalent planes each normal to the appropriate spring then the total force vector is the sum of the individual radial forces and the vertical and fore and aft components of force at the axle are obtained.

1.28 TYRE VIBRATION MODES

Vibration modes involving the movement of the tread in a radial manner have been noted, some of these modes are not stationary and involve the nodes rotating around the periphery in either clock or counter clockwise direction. Longitudinal vibrations are also present. The radial tyre has a number of sharply defined resonances while the bias ply tyre has a high background level but less pronounced resonances.

Test procedures may consist of vibrating the tyre footprint while the tyre is stationary or of running the tyre on a drum or flat belt embossed with a representative road profile. The results obtained from either method appear comparable.

The differences between the sidewall and tread band for the radial construction which give the tread a greater stiffness and mass while the radial flexibility is contained in the sidewalls accounts for the fact that vibration tests on these tyres show distinct resonances while the more homogeneous bias ply construction has a response in which the resonances are less marked but the background level is higher.

As with many other aspects of tyres a mechanical model has been developed to study vibration modes. In this case the model is that of a cylinder with tension and bending stiffness in the radial direction supported on a continuous spring foundation. It is not proposed to develop the analysis for this particular model but it is interesting to note that it shows that the typical mode shapes of the vibration tyre need not be stationary but can travel around the tyre periphery either in the direction of rotation or in the opposite sense. The sketch of the computed mode shapes for a radial tyre is given in Fig.1.28, together with values of measured frequencies.

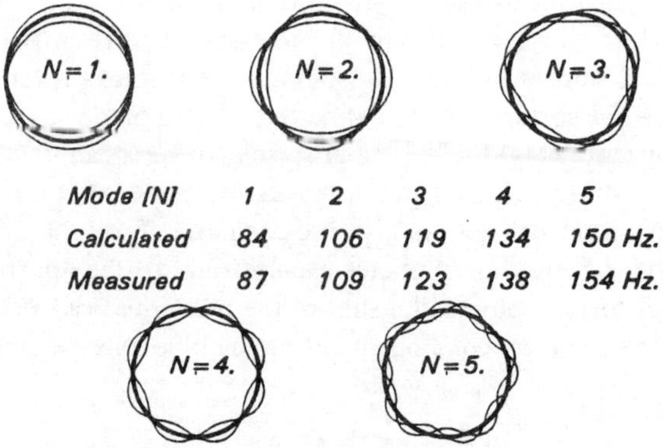

Mode [N]	1	2	3	4	5
Calculated	84	106	119	134	150 Hz.
Measured	87	109	123	138	154 Hz.

Fig.1.28 Vibration modes of a rotating radial tyre

For the purpose of simulating the response of a tyre to road roughness the modes of Fig.1.28 can be represented by a parallel set of spring, mass, damper systems set so that one of the masses resonates at each of the measured, or computed, modes. When the parallel second order systems are subjected to a random input then each will provide an input to the wheel hub corresponding to the particular frequency and these inputs will be summed in amplitude and phase at the axle. Although the dampers are not shown in the sketch they have an important role to play in limiting the peak values and in providing the background 'noise'.

1.29 COMMENT

This chapter has presented the tyre as a device which responds to stimulation. For example the application of a force produces deflection. Steering the rolling tyre causes the tread band to move laterally relative to the wheel rim which produces a lateral force and, since the deflection is not uniform, a moment is also generated. The purpose of the first part of the text is to demonstrate the logical manner in which the tyre responds to various demands.

The tyre models shown are essentially methods of fitting, or approximating those tyre characteristics used in vehicle ride and handling models. They do not provide a basis from which tyre design can be assessed with regard to ride and handling performance. When the tyre is in service it will meet conditions not experienced on the test stand and the data must be modified empirically to suit the new situations.

The ride characteristics of tyres have not received as much attention as the steering responses. In general the tyre is still conceived as a vertical spring with a small amount of damping, or as a Maxwell spring system. The text includes a diagram of measured mode shapes for tyre vibrations and a note of the effect of short wavelength irregularities.

A bibliography of most of the tyre studies before 1970 is contained in the first two references. The first of these is the work of one person while the second is a set of dissertations on 'the state of the art' by a number of authors. The reader is advised to use the original texts given in these references.

REFERENCES

(1) R.HADEKEL (1950) *The mechanical characteristics of pneumatic tyres*, UK Ministry of Supply, TPA 3.

(2) S.K.CLARK (Editor) (1971) *Mechanics of pneumatic tires*, US National Bureau of Standards, Monograph 122.

(3) J.R.ELLIS and F.FRANK (1966) *The equilibrium shapes of tyres*, Cranfield Institute of Technology, ASAE Report No.1.

(4) W.J.MORLAND (1951) *Landing gear vibration*, AF TR 6590.

(5) J.S.LOEB (1984) *Tire lateral stiffness and its effect on relaxation length*, MSc Dissertation, Ohio State University.

(6) D.C.DAVIS (1979) 'A radial spring terrain enveloping tyre model', *Vehicle System Dynamics*, 3(1).

(7) K.M.CAPTAIN, A.B.BOGHANI, and D.N.WORMLEY (1979) 'Analytical tyre models for dynamic vehicle simulations', *Vehicle System Dynamics*, 8(1).

(8) E.BAKKER, L.Nyborg, and H.B.PACEJKA (19??) *Tire modelling for use in vehicle dynamics studies*, SAE 870421.

CHAPTER TWO

Axis Systems and Equations of Motion

2.1 AXIS SYSTEM

The equations of motion employed in vehicle studies are related to a set of body-fixed axes, this axis system is right handed. The positive sense of rotation is clockwise when viewed from the origin along the positive direction of the axis. Figure 2.1 shows the system. The origin of the axes is usually located at the centre of mass of the total vehicle in the case of cars. The fifth wheel is a convenient location for the origin of both tractor and semi-trailer axes in the case of an articulated semi-trailer vehicle.

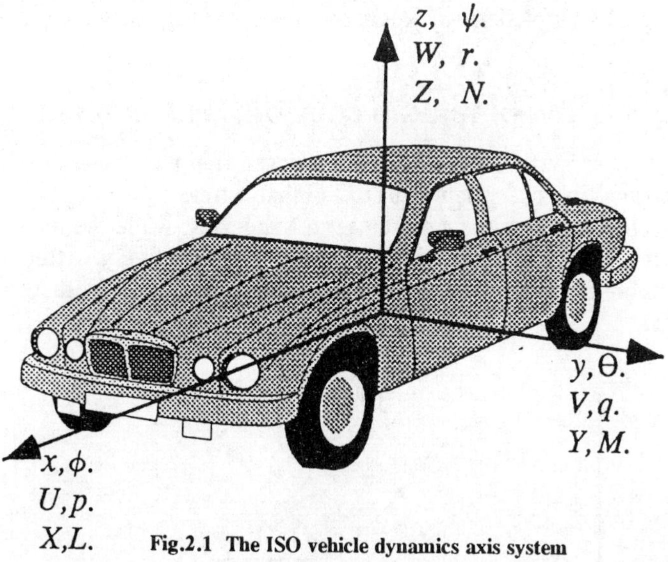

Fig.2.1 The ISO vehicle dynamics axis system

Road vehicles are controlled by forces and moments developed at the interface between tyre and road. The nature of these forces and moments is such that it is convenient to consider the equations of motion in two phases. The lateral and yawing motions cause the tires to generate angles of relative velocity against the road, thus the equations in sideslip and yaw are first order equations.

Roll, bounce, and pitch movements call into play the springs and dampers of the suspension which act in series with the tyre springs in addition to the slip angle properties and thus the equations of motion are of second order.

It will be seen that the slip and steer angles of the tires contain terms due to both the velocities and movements of roll pitch and bounce. Although there is no intrinsic difficulty in formulating sets of equations which consider all possible motions this exercise is not usually required since it is possible to explain vehicle performance by using three sub-sets.

(1) Forward speed, lateral velocity, yaw velocity, and roll angle.
(2) Bounce, pitch, and roll.
(3) Forward speed, bounce, and pitch.

In the first subset the forward speed, lateral velocity and yaw velocity form a closely coupled set of equations which is frequently considered on its own.

2.2 MOTION OF A RIGID BODY ON A PLANE SURFACE

These equations will be developed from first principles. Consider the case of a body moving on a plane surface so that a time $t = 0$ the body fixed axis set is inclined at angle ϕ to the earth fixed axes, while the longitudinal lateral and yawing velocities are U, V, and r, respectively. After time δt each variable has increased by a small amount. Figure 2.2 illustrates the condition.

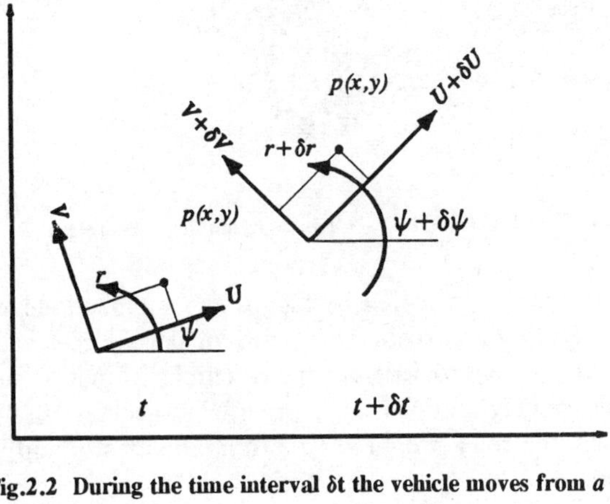

Fig.2.2 During the time interval δt the vehicle moves from *a* to *b*

At time $t = 0$, the velocities of point $P(x,y)$ fixed in the body are determined.

$$u = U - yr$$

$$v = V + xr \tag{2.1}$$

At time $t = \delta t$ the body has moved to the second position.

$$u' = (U+\delta U) - y(r+\delta r)$$

$$v' = (V+\delta V) + x(r+\delta r) \tag{2.2}$$

Resolve the new velocities along the original directions and subtract the original values

$$u = u'\cos\delta\psi - v'\sin\delta\psi - u$$

$$v = u'\sin\delta\psi + v'\cos\delta\psi - v \tag{2.3}$$

Substitute the appropriate values for u, v, r, etc.

$$u = \{(U+\delta U)-y(r+\delta r)\}\cos\delta\psi$$
$$- \{(V+\delta V)+x(r+\delta r)\}\sin\delta\psi - (U-yr)$$

$$v = \{(U+\delta U)-y(r+\delta r)\}\sin\delta\psi$$
$$+ \{(V+\delta V)+x(r+\delta r)\}\cos\delta\psi - (V+xr)$$

As δt becomes small $\cos\delta\psi \to 1$ and $\sin\delta\psi \to \delta\psi$. Substitute these approximations and divide by δt

$$\delta u/\delta t = \delta U/\delta t - y\delta r/\delta t - V\delta\psi/\delta t - \delta V\delta\psi/\delta t$$
$$- xr\delta\psi/\delta t - x\delta r\delta\psi/\delta t$$

$$\delta v/\delta t = \delta V/\delta t + x\delta r/\delta t + U\delta\psi/\delta t + \delta U\delta\psi/\delta t$$
$$- yr\delta\psi/\delta t - y\delta r\delta\psi/\delta t$$

Let $\delta t \to 0$, then $\delta(\,)/\delta t \to d(\,)/dt$; also $d\delta\psi/dt = r$; $\delta U, \delta V, \delta r \to 0$.

$$a(x) = \dot{u} = \dot{U} - Vr - y\dot{r} - xr^2$$

$$a(y) = \dot{v} = \dot{V} + Ur + x\dot{r} - yr^2 \tag{2.4}$$

The accelerations of point $P(x,y)$ are given by equation (2.4). Let this point represent an element of mass, δm. The summation of these elements in the x and y directions and the sum of their moments around the z axis equal the applied forces and moment, respectively, as illustrated in Fig.2.3.

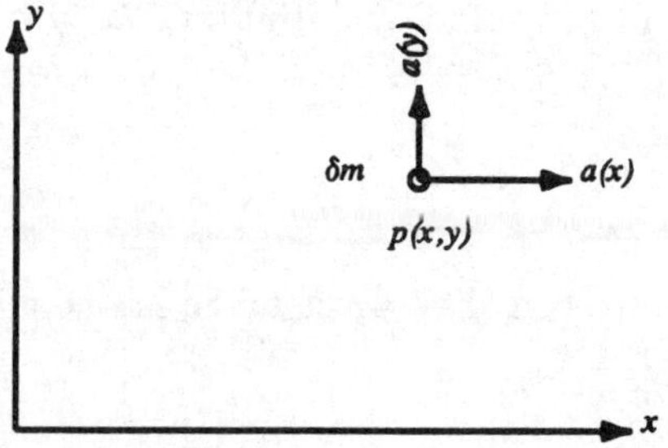

Fig.2.3 The accelerations of a typical element of mass

$$\Sigma X = \Sigma \delta m a(x)$$
$$\Sigma Y = \Sigma \delta m a(y) \qquad (2.5)$$
$$\Sigma N = \Sigma \delta m \{x.a(y) - y.a(x)\}$$

Substitute for the expressions $a(x)$ and $a(y)$

$$\Sigma X = \Sigma \delta m (U\text{-}Vr\text{-}yr\text{-}xr^2)$$
$$\Sigma Y = \Sigma \delta m (V+Ur+xr\text{-}yr^2) \qquad (2.6)$$
$$\Sigma N = \Sigma \delta m \{x.(V+Ur+xr\text{-}yr^2) - y.(U\text{-}Vr\text{-}yr\text{-}xr^2)\}$$

$$\Sigma \delta m = m;\ \ \Sigma \delta m x = m\bar{x}\ ;\ \ \Sigma \delta m y = m\bar{y}$$

Also

$$\Sigma \delta m (x^2+y^2) = I_z$$

Hence

$$\Sigma X = m(\dot{U}\text{-}Vr) - m\bar{y}\dot{r} - m\bar{x}r^2$$
$$\Sigma Y = m(\dot{V}+Ur) + m\bar{y}\dot{r} - m\bar{x}r^2 \qquad (2.7)$$
$$\Sigma N = I_z\dot{r} + m\bar{x}(\dot{V}+Ur) - m\bar{y}(U\text{-}Vr)$$

When the origin is located at the centre of gravity of the body the equations of motion are greatly simplified.

$$\Sigma X = m(\dot{U}\text{-}Vr)$$
$$\Sigma Y = m(\dot{V}+Ur) \qquad (2.8)$$
$$\Sigma N = I_z\dot{r}$$

The development of the forces and moment by the action of the tires will be discussed later.

2.3 THE VEHICLE PATH

The equations for the body centered axis system provide information about the linear and angular velocities of the body. There are many instances where the position on the road is significant and this may be achieved by open loop integration of the velocities.

$$\psi = \psi_0 + \int r dt$$

$$x(path) = \int (U\cos\psi - V\sin\psi)dt \qquad (2.9)$$

$$y(path) = \int (U\sin\psi + V\cos\psi)dt$$

2.4 MOTION OF A SIX-DEGREES-OF-FREEDOM BODY

The classic presentation of a body moving in space is written in terms of the linear and angular velocities. When used in vehicle simulations these equations give rise to problems associated with the open loop integration of velocities to establish the displacements of the suspension and tyre springs and the stability of the simulation is dependent upon numerical truncation and noise.

The axis system is right handed. The positive rotation is clockwise around the axis when viewed from the origin along the positive direction.

U, U, W are the instantaneous velocities of the origin in the direction of the axis at that instant of time. The angular velocities are p,q,r around the x,y,z axes respectively. The point $P(x,y,z)$ is a fixed point relative to the moving axes.

A is the position vector.
V is the velocity vector.
ω is the angular velocity vector.
a is the acceleration vector.

The velocity of P is $V = V(0) + \omega \times A$

$$V(0) = \begin{bmatrix} U & i \\ V & j \\ W & k \end{bmatrix}$$

$$\omega = \begin{bmatrix} p & i \\ q & j \\ r & k \end{bmatrix}$$

$$A = \begin{bmatrix} x & \\ y & j \\ z & k \end{bmatrix}$$

$$V = \begin{bmatrix} U \\ V \\ W \end{bmatrix} \begin{bmatrix} i \\ j \\ k \end{bmatrix} + \begin{bmatrix} i & j & k \\ p & q & r \\ x & y & z \end{bmatrix}$$

$$V = \begin{bmatrix} v(x) \\ v(y) \\ v(z) \end{bmatrix} \begin{bmatrix} i \\ j \\ k \end{bmatrix} = \begin{bmatrix} U - yr + zq \\ V - zp + xr \\ W - xq + yp \end{bmatrix} \begin{bmatrix} i \\ j \\ k \end{bmatrix}$$

Differentiation of this equation gives the acceleration components of $P(x,y,z)$ relative to the body fixed origin.

$$a(x0) = \dot{U} - yr + zq$$

$$a(y0) = \dot{V} - zp + xr$$

$$a(z0) = \dot{W} - xq + yp$$

The acceleration of P relative to a ground fixed origin is

$$a = a(0) + \omega \times V$$

$$a = \begin{bmatrix} a(x0) \\ a(y0) \\ a(z0) \end{bmatrix} \begin{bmatrix} i \\ j \\ k \end{bmatrix} + \begin{bmatrix} i & j & k \\ p & q & r \\ v(x) & v(y) & v(z) \end{bmatrix}$$

Hence

$$a = \begin{bmatrix} a(x0) + qv(z) - rv(y) \\ a(y0) + rv(x) - pv(z) \\ a(z0) + pv(y) - qv(x) \end{bmatrix} \begin{bmatrix} i \\ j \\ k \end{bmatrix}$$

Substitute the component velocities of *P*. Then

$$a(x) = \dot{U} - Vr + Wq - (q^2+r^2)x + (pq-\dot{r})y + (pr+\dot{q})z$$

$$a(y) = \dot{V} - Wp + Ur - (p^2+r^2)y + (qr-\dot{p})z + (pq+\dot{r})x \qquad (2.10)$$

$$a(z) = \dot{W} - Uq + Vp - (p^2+q^2)z + (pr-\dot{q})x + (qr+\dot{p})y$$

Thus the velocity and acceleration of a point rigidly attached to a body moving with six degrees of freedom are defined for the condition that the reference axes are fixed in the moving body. The inertial equations are obtained from a summation of the effects of the small elements of mass.

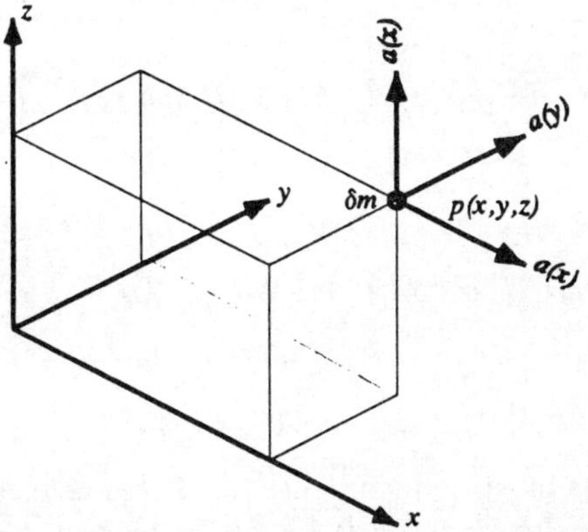

Fig.2.4 Acceleration of a typical element of mass in three dimensions

The linear equations are

$$\Sigma X = \Sigma \delta m . a(x)$$

$$\Sigma Y = \Sigma \delta m . a(y)$$

$$\Sigma Z = \Sigma \delta m . a(z)$$

The mass and mass-moments are

$$\Sigma \delta m = m; \;\; \Sigma \delta m x = m\bar{x}; \;\; \Sigma \delta m y = m\bar{y}; \;\; \Sigma \delta m z = m\bar{z}$$

$$\Sigma X = m(\dot{U} - Vr + Wq) - m\bar{x}(q^2+r^2) + m\bar{y}(pq-\dot{r}) + m\bar{z}(pr+\dot{q})$$

$$\Sigma Y = m(\dot{V} - Wp + Ur) + m\bar{x}(pq+\dot{r}) - m\bar{y}(p^2+r^2) + m\bar{z}(qr-\dot{p})$$

$$\Sigma Z = m(\dot{W} - Uq + Vp) + m\bar{x}(pr-\dot{q}) + m\bar{y}(qr+\dot{p}) - m\bar{z}(p^2+q^2)$$

The moments around the axes are

$$\Sigma L = \Sigma\delta m\{a(z)y - a(y)z\}$$

$$\Sigma M = \Sigma\delta m\{a(x)z - a(z)x\}$$

$$\Sigma N = \Sigma\delta m\{a(y)x - a(x)y\}$$

The moments and products of inertia are

$$I_x = \Sigma\delta m(y^2+z^2); \qquad P_{xy} = \Sigma\delta m(xy)$$

$$I_y = \Sigma\delta m(x^2+z^2); \qquad P_{xz} = \Sigma\delta m(xz)$$

$$I_z = \Sigma\delta m(x^2+y^2); \qquad P_{yz} = \Sigma\delta m(yz)$$

Then

$$\begin{aligned} \Sigma L &= I_x\dot{p} - (I_y - I_z)qr + P_{yz}(r^2-q^2) - P_{xz}(pq+\dot{r}) + P_{xy}(pr-\dot{q}) \\ &\quad + m\bar{y}(\dot{W}-Uq+Vp) - m\bar{z}(\dot{V}-Wp+Ur) \\ \Sigma M &= I_y\dot{q} - (I_z - I_x)pr + P_{xz}(p^2-r^2) - P_{xy}(qr+\dot{p}) + P_{yz}(qp-\dot{r}) \\ &\quad + m\bar{z}(\dot{U}-Vr+Wq) - m\bar{x}(\dot{W}-Uq+Vp) \\ \Sigma N &= I_z\dot{r} - (I_x - I_y)pq + P_{xy}(q^2-p^2) - P_{yz}(pr+\dot{q}) + P_{xz}(qr-\dot{p}) \\ &\quad + m\bar{x}(\dot{V}-Wp+Ur) - m\bar{y}(\dot{U}-Vr+Wq) \end{aligned} \tag{2.11}$$

The car models most frequently used will assume that symmetry exists in the xz plane, thus $m\bar{y} = 0$ and $P_{xy} = 0$. The models also assumes that the car is operating on a smooth surface so that the dynamic effects of pitch and bounce, the ride modes, can be ignored.

$$\begin{aligned} \Sigma X &= m(\dot{U}-Vr) - m\bar{x}r^2 + m\bar{z}pr && \text{[fore/aft]} \\ \Sigma Y &= m(\dot{V}+Ur) - m\bar{x}\dot{p} + m\bar{z}\dot{r} && \text{[lateral]} \\ \Sigma L &= I_x\dot{p} + P_{yz}r^2 - P_{xz}\dot{r} - m\bar{z}(\dot{V}+Ur && \text{[yaw]} \\ \Sigma N &= I_z\dot{r} - P_{yz}pr - P_{xz}\dot{p} + m\bar{x}(\dot{V}+Ur) && \text{[roll]} \end{aligned} \tag{2.12}$$

If it is further assumed that the origin is at the centre of gravity and that the axes are the principal axes of inertia then the most simple set of equations is obtained.

$$\Sigma X = m(\dot{U}-Vr)$$

$$\Sigma Y = m(\dot{V}+Ur)$$

$$\Sigma L = I_x \dot{p}$$

$$\Sigma N = I_z \dot{r}$$

Articulated vehicle models will be compiled with the origin at the fifth wheel coupling between tractor and trailer.

2.5 THE ROTATION MATRIX

The need to calculate the positions of the centers of contact and the relative velocities of the tyre contact points so that the tyre force and moments can be referred to the inertial axes is satisfied by using a rotation matrix.

The rotations are not unique and, for given values of the angles, the calculated position of a typical point *P(x,y,z)* fixed in the rotating axis system depends upon the order of rotations. In this text the order of rotations is yaw (rotation around the *z* axis), pitch (rotation around the displaced *y* axis, y_1) and roll (rotation around the *x* axis, itself displaced to x_1 by the yawing movement and then to x_2 by the pitch). Note that while the angles must be treated as non commutable the angular velocities do not have this restriction. The advantage of the the fact that angular velocities are a linear set is frequently employed in simulation.

The rotation around the z_0 axis gives the first intermediate position of *P*.

$$\begin{bmatrix} \cos\psi & -\sin\psi & 0 \\ \sin\psi & \cos\psi & 0 \\ 0 & 0 & 1 \end{bmatrix} \begin{bmatrix} x_1 \\ y_1 \\ z_1 \end{bmatrix} = \begin{bmatrix} x_0 \\ y_0 \\ z_0 \end{bmatrix}$$

The second rotation is around the displaced pitch axis y_1 and gives the second position of *P*.

$$\begin{bmatrix} \cos\Theta & 0 & \sin\Theta \\ 0 & 1 & 0 \\ -\sin\Theta & 0 & \cos\Theta \end{bmatrix} \begin{bmatrix} x_2 \\ y_2 \\ z_2 \end{bmatrix} = \begin{bmatrix} x_1 \\ y_1 \\ z_1 \end{bmatrix}$$

The final rotation is around the roll axis which has been displaced from its initial position by the two previous rotations.

$$\begin{bmatrix} 1 & 0 & 0 \\ 0 & \cos\phi & -\sin\phi \\ 0 & \sin\phi & \cos\phi \end{bmatrix} \begin{bmatrix} x_3 \\ y_3 \\ z_3 \end{bmatrix} = \begin{bmatrix} x_2 \\ y_2 \\ z_2 \end{bmatrix}$$

Substitution of the intermediate values gives the final relationship.

$$\begin{bmatrix} \cos\Theta\cos\psi & \sin\Theta\sin\phi\cos\psi - \cos\phi\sin\psi & \sin\Theta\cos\phi\cos\psi + \sin\phi\sin\psi \\ \cos\Theta\sin\psi & \sin\Theta\sin\phi\sin\psi + \cos\phi\cos\psi & \sin\Theta\cos\phi\sin\psi - \sin\phi\cos\psi \\ -\sin\Theta & \cos\Theta\sin\phi & \cos\Theta\cos\phi \end{bmatrix} \begin{bmatrix} x_3 \\ y_3 \\ z_3 \end{bmatrix} = \begin{bmatrix} x_0 \\ y_0 \\ z_0 \end{bmatrix}$$

Note that since the inverse of this matrix is also its transform the relation from known ground coordinates to the body fixed coordinate may be obtained by interchanging rows and columns in the matrix.

2.6 MOMENTS AND PRODUCTS OF INERTIA

The moments and products of inertia of a body are defined about a set of mutually perpendicular axes which can be selected in an arbitrary manner. There will be one set of axes for which all the products of inertia are zero, the principal axes of inertia for the body. If the body possesses a plane of symmetry and two axes are in this plane, (e.g., x,y); then for every element of mass at a positive value of z an opposing element exists at $-z$ from the plane of symmetry. If two mutually perpendicular planes of symmetry exist and the axes are selected to be in these planes then all products of inertia are zero. The axes so selected are the principal axes of inertia for the body and the corresponding moments of inertia are the principal moments of inertia.

Parallel axes

When the moments and products of inertia of a body are known for a particular set of axes the corresponding values for a parallel axis set may be calculated.

Let the first axis set have the origin at the centre of gravity of the body. I_x, I_y, I_z, P_{xy}, P_{xz}, P_{yz} are the known values.

The new axis set is parallel and distant x', y', z' from the original set.

$$I_{x'} = \Sigma\delta m\{(y+y')^2 + (z+z')^2\}$$

$$I_{x'} = \Sigma\delta m(y^2+z^2)+\Sigma\delta m(y'^2+z'^2)+2y'\Sigma\delta my+2z'\Sigma\delta mz$$

But

$$m\bar{z} = m\bar{y} = 0$$

Thus

$$I_{x'} = I_x + m(y'^2+z'^2) \tag{2.13}$$

The product of inertia terms is

$$P_{x'y'} = \Sigma\delta m(x+x')(y+y')$$

or

$$P_{x'y'} = P_{xy} + mx'y' \tag{2.14}$$

Similar expressions may be developed for the other moments and products of inertia.

Skew axes

The other transformation required is that for a skew axis set through the centre of gravity. First consider the direction cosines relating the axes.

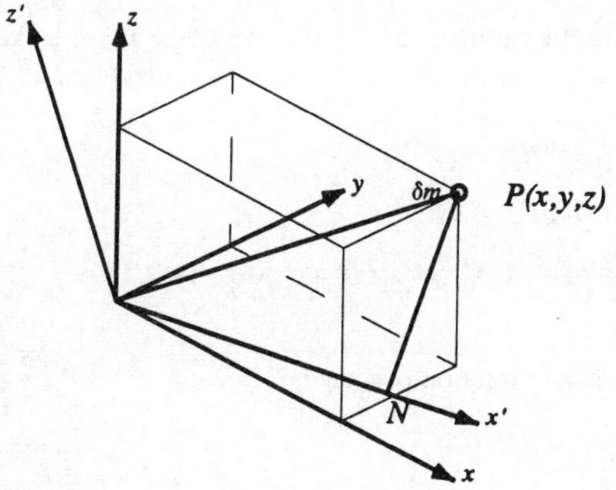

Fig.2.5 The original inertial axes and a skewed set of axes. A rotation around the y axis is shown in this diagram

$$x' = a_1x + b_1y + c_1z$$

$$y' = a_2x + b_2y + c_2z$$

$$z' = a_3x + b_3y + c_3z$$

where

$$a_1^2 + b_1^2 + c_1^2 = 1$$

$$a_2^2 + b_2^2 + c_2^2 = 1$$

$$a_3^2 + b_3^2 + c_3^2 = 1$$

also

$$a_1^2 + a_2^2 + a_3^2 = 1$$

$$b_1^2 + b_2^2 + b_3^2 = 1$$

$$c_1^2 + c_2^2 + c_3^2 = 1$$

Since the x and y axes are perpendicular

$$b_1c_1 + b_2c_2 + b_3c_3 = 0$$

$$a_1c_1 + a_2c_2 + a_3c_3 = 0$$

$$a_1b_1 + a_2b_2 + a_3b_3 = 0$$

To determine the moment of inertia of a body about a skew axis $O_{x'}$ in terms of the known values of the moments and products of inertia for the original axes.

$$I_x' = \Sigma\delta m.PN^2$$

$$I_x' = \Sigma\delta m(OP^2 - ON^2)$$

$$I_x' = \Sigma\delta m\{(x^2 + y^2 + z^2) - (a_1x + b_1y + c_1z)^2\}$$

A property of the direction cosines is

$$a_1^2 + b_1^2 + c_1^2 = 1$$

Thus

$$I_{x'} = a_1^2I_x + b_1^2I_y + c_1^2I_z - 2(b_1c_1P_{yz} + a_1b_1P_{xy} + a_1c_1P_{xz})$$

The product of inertia about the axes $O_{x'}, O_{y'}$ is

$$P_{x'y'} = \Sigma \delta m (a_1 x + b_1 y + c_1 z)(a_2 x + b_2 y + c_2 z)$$

$$
\begin{aligned}
P_{x'y'} = {} & -a_1 a_2 I_x - b_1 b_2 I_y - c_1 c_2 I_z \\
& + Pxy(a_1 b_2 + a_2 b_1) \\
& + Pxz(a_1 c_2 + a_2 c_1) \\
& + Pyz(b_1 c_2 + b_2 c_1)
\end{aligned}
$$

With these formulae test data can be transformed to any desired axis set.

2.7 STABILITY OF SYSTEMS

A system is stable if, following a disturbance it moves to a position of finite equilibrium. From the point of view of a vehicle operated by a driver the statement is not sufficient, since the vehicle must also be seen to respond so that it remains within the bounds of the highway and at a rate which does not cause the driver to over control. Driver-induced instability is always a possibility. The following discussion is concerned only with the vehicle itself.

Stability studies are carried out by considering the effect of a small perturbation about a condition of equilibrium. During this small motion the equations of motion of the vehicle will be linear differential equations with constant coefficients. In such a system each part of the motion will conform to a type $\exp(\sigma t)$ and the stability of the system is considered in terms of the coefficient .Thus if σ is real and positive the system is unstable since the amplitude will increase with time, a divergent motion. A real and negative value of σ indicates that the system will converge to some steady state value, and thus will be stable. A complex value of σ with a positive real part indicates the presence of a divergent oscillatory motion while a negative real part shows that a decaying oscillation is present. The presence of a positive real root will inevitably lead to divergence no matter how small that real value. Thus the stability ofa system is demonstrated by examination of the roots of the characteristic equation.

A set of criteria for stability have been determined by Routh and are usually known as the Routh-Hurwicz criteria, an outline of the procedure follows. The characteristic frequency equation is written in polynomial form.

$$A_n D^n + A_{n-1} D^{n-1} + \ldots + A_{n-1} D + A_0 = 0$$

A necessary but not sufficient condition for stability is that all coefficients shall be positive. The complete requirements for stability are satisfied if the following test functions are all positive.

$$T_1 = A_{n-1}$$

$$T_2 = \begin{vmatrix} A_{n-1} & A_n \\ A_{n-3} & A_{n-2} \end{vmatrix}$$

$$T_3 = \begin{vmatrix} A_{n-1} & A_n & 0 \\ A_{n-3} & A_{n-2} & A_{n-1} \\ A_{n-5} & A_{n-4} & A_{n-3} \end{vmatrix}$$

$$T_4 = \begin{vmatrix} A_{n-1} & A_n & A_0 & A_0 \\ A_{n-3} & A_{n-2} & A_{n-1} & A_n \\ A_{n-5} & A_{n-4} & A_{n-3} & A_{n-2} \\ A_{n-7} & A_{n-6} & A_{n-5} & A_{n-4} \end{vmatrix}$$

Routh also demonstrated that the number of roots with real parts positive in an unstable system is equal to the number of sign changes in the polynomial.

The quartic is of particular interest in vehicle dynamics.

$$A_4D^4 + A_3D^3 + A_2D^2 + A_1D + A_0 = 0$$

For a stable system the following conditions must be satisfied.

$$T_1 = A_3 > 0$$

$$T_2 = \begin{vmatrix} A_3 & A_4 \\ A_1 & A_2 \end{vmatrix} > 0$$

$$T_3 = \begin{vmatrix} A_3 & A_4 & 0 \\ A_1 & A_2 & A_3 \\ 0 & A_0 & A_1 \end{vmatrix} > 0$$

Hence

$$T_3 = A_1T_2 - A_0A_3^2$$

If A_4 is made equal to unity, then the quartic may be expressed

$$D^4 + A_3D^3 + A_2D^2 + A_1D + A_0 = 0$$

Then, for stability

$$A_2 > 0; \quad A_0 > 0; \quad A_3A_2A_1 - A_0A_3^2 - A_1^2 > 0$$

2.8 CRITICAL STABILITY CRITERIA

If a system is known to be stable for some standard configuration then it is possible to assess the stability in a different loading state, for example, without considering all the inequalities in detail. This method is useful when the effect of varying one parameter is needed.

In the stable condition all the roots of the characteristic equation are negative or have negative real parts. The potential for instability requires that either a real negative root becomes zero, or the real part of a pair of complex roots becomes zero.

The change of sign of a real root in an otherwise stable system occurs when:

$$A_0 = 0$$

Complex roots occur in pairs; hence, in the critical case the roots are $\pm$ip. A quartic equation is used to demonstrate the Routh - Hurwicz criteria for the case of complex roots, that is when an oscillatory response becomes undamped.

$$A_4D^4 + A_3D^3 + A_2D^2 + A_1D + A_0 = 0$$

For the critical condition the value $\pm$ip is substituted for D.

$$A_4p^4 +/- iA_3p^3 + A_2p^2 +/- iA_1p + A_0 = 0$$

Equate the real and imaginary parts of this equation.

Real: $A_4p^4 + A_2p^2 + A_0 = 0$

Imaginary: $A_3p^3 + A_1p = 0$

or

$$p^2 = -A_1/A_3$$

Substitute for p^2 in the equation of the real parts.

$$A_3A_2A_1 - A_0A_3^2 - A_1^2 = 0$$

Which is

$$T_3 = 0$$

This argument can be extended to show that in any polynomial the criterion for a pair of imaginary roots is that the penultimate test function T_{n-1} be zero.

CHAPTER THREE

The Control and Stability of Basic Rigid Vehicles

A basic vehicle is considered to be one which uses the variables of speed, lateral velocity and yawing velocity to describe the behaviour resulting from disturbances due to steering, acceleration, braking, and wind gusts.

The equations of motion have been developed in Chapter Two and those variables are used to develop the slip angles at the tyre/road contact surfaces. Steering is an input to the equations which thus develop transient and steady state phases.

3.1 DEFINITIONS

Some basic characteristics of vehicles which will be met in this chapter are brought together here for emphasis.

Ideal (Ackerman) angles. The steering and side-slip angles which occur when a vehicle is driven slowly around a constant radius turn.

Understeer. An understeering vehicle will, when tested on a constant radius turn, require the steering angle to be increased as speed is increased.

Oversteer. An oversteering vehicle will, on a constant radius turn, require the steering angle to be reduced as speed is increased.

Neutral steer. A neutral steering vehicle will, on a constant radius turn, maintain a constant steer angle independent of speed.

Steering characteristic (Gain). The slope of the plot of steer angle versus lateral acceleration for a constant radius test (ds/dG).

Characteristic speed. The speed at which an understeering vehicle has its greatest steady state yaw velocity gain ($r/s \mid_{ss}$).

Critical speed. The speed at which an oversteering vehicle becomes unstable.

Neutral steer point. That position away from the front axle of a vehicle at which, theoretically, an applied lateral force causes the vehicle to side-slip without changing heading direction.

Static margin. The distance between the neutral steer point and the centre of gravity of a vehicle. Using the definition of tyre characteristics used in this text a negative static margin indicates an understeering vehicle.

3.2 TYRE ATTITUDE ANGLES AND VEHICLE VARIABLES

The *attitude angle* for a tyre has been defined in section 1.3. An expression for each tyre is obtained in terms of the variables; speed (U), lateral velocity (V), and yaw velocity (r). Figure 3.1 is a plan view of a basic car in which the centre of gravity is located at a behind the front axle and b ahead of the rear axle. The wheelbase is L ($=a+b$). The longitudinal and lateral velocities at the centres of tyre/road contact specify the slip angle at each road wheel.

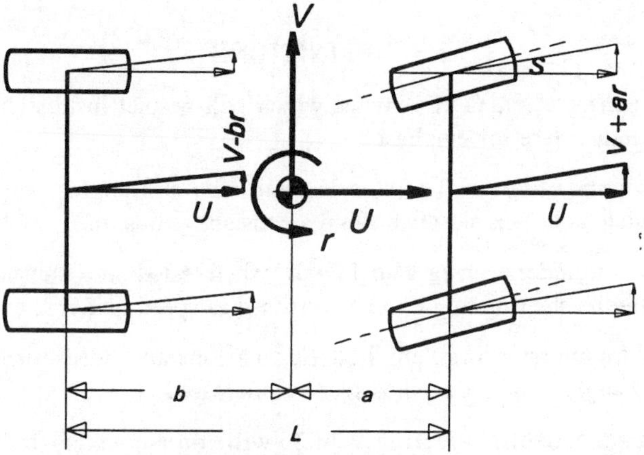

Fig.3.1 The attitude angles at the wheels are replaced by mean values for the pair of wheels in the 'bicycle' model

Although the slip angles for the left and right wheels on an axle are slightly different this difference is unimportant in most cases because the yawing velocity effect on the local forward speed at the contact patch is of low order. A significantly high yaw velocity indicates that the vehicle is out of control.

For all practical purposes the attitude angles are small and thus the angle in radians is equivalent to the arctangent.

From Fig.3.1 it can be seen that the front wheels are steered relative to the frame through an angle, s. The basic slip angles for the front and rear axles are the vector velocities shown.

	Rear Axle	Front Axle
Steer		s
Slip	$(V - br)/U$	$(V + ar)/U$
Attitude	$-(V - br)/U$	$s - (V + ar)/U$

3.3 TWO-DEGREES-OF-FREEDOM CAR MODEL, EQUATIONS OF MOTION

This model is capable of demonstrating a number of important basic facts of vehicle chassis design and the effects of road speed, wind disturbances and road camber. The basic car operates at constant forward speed with lateral velocity and yawing velocity as the two variables of the body-centred axis set. For a body moving at constant forward speed on a plane surface the equations of motion are given below.

$$\begin{aligned} \Sigma X &= m(\dot{V} + Ur) = Y_1 + Y_2 \\ \Sigma N &= I_z\dot{r} = aY_1 - bY_2 + N_1 + N_2 \end{aligned} \tag{3.1}$$

The justification for ignoring the fore/aft inertial equation is that there is no inertial coupling between this mode and the side-slip/yaw modes.The effects of acceleration/braking on handling are mainly through the changes induced in the tyre characteristics. Figure 3.2 shows a car moving forward with side-slip and yawing velocities.

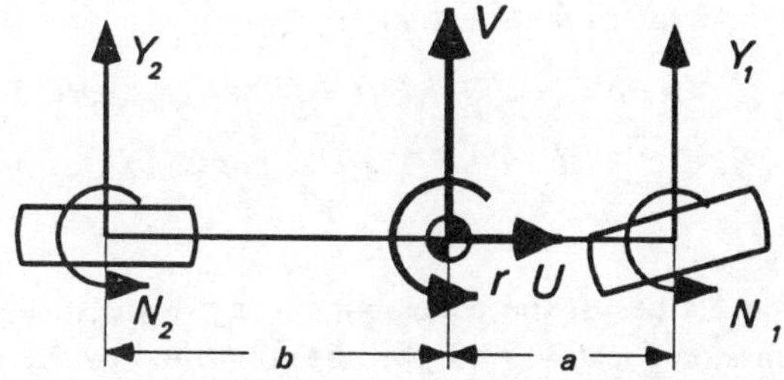

Fig.3.2. Basic handling model with 2 degrees of freedom; lateral velocity (V) and yawing velocity (r); speed (U) is constant

Tyre lateral forces and aligning moments are due to the attitude angles developed at front and rear axles.

For small attitude angles these force and moment relations are linear. Both wheels act together and the coefficient C defines the initial slope of the lateral force versus steer angle curve for each axle. $\hat{T}$ is the slope of the aligning moment curve for an axle.

Let

$$C_1 = 2*dY/ds_1; \quad C_2 = 2*dY/ds_2$$
$$\hat{T}_1 = 2*dN/ds_1; \quad \hat{T}_2 = 2*dN/ds_2$$

$$\begin{aligned} m(\dot{V}+Ur) &= C_1\{s-(V+ar)/U\} + C_2\{-(V-br)/U)\} \\ I_z\dot{r} &= (\hat{T}_1+aC_1)\{s-(V+ar)/U\} + (\hat{T}_2-bC_2)\{-(V-br)/U)\} \end{aligned} \tag{3.2}$$

The Heaviside notation $D(\ ) = d(\ )/dt$ is used, and the equations are written in matrix form.

$$\begin{bmatrix} mD+(C_1+C_2)/U & mU+(aC_1-bC_2)/U \\ (aC_1-bC_2+\hat{T}_1+\hat{T}_2)/U & ID+(a^2C_1+b^2C_2 \\ & +a\hat{T}_1-b\hat{T}_2)/U \end{bmatrix} \begin{bmatrix} V \\ r \end{bmatrix} = \begin{bmatrix} C_1 \\ aC_1 \end{bmatrix} s$$

3.4 TWO-DEGREE-OF-FREEDOM CAR; CRITICAL/CHARACTERISTIC SPEED

The characteristic equation of the system is the left hand square matrix of equation (3.3). This is expanded into polynomial form in order to study the frequency and damping of the equation.

$$\begin{aligned} & D^2 + \{(a^2C_1+b^2C_2+a\hat{T}_1+b\hat{T}_2)/I_z+(C_1+C_2)/m\}D/U \\ & + [\{L^2C_1C_2+L(C_1\hat{T}_2+C_2\hat{T}_1)\}/U^2 - m(aC_1-bC_2+\hat{T}_1+\hat{T}_2)]\, D^0/(mI_z) \\ & = 0 \end{aligned} \tag{3.3}$$

The physical values of the terms containing $\hat{T}$ are small and for the following discussion equation (3.3) will be simplified by the elimination of such terms.

$$D^2 + \{(a^2C_1+b^2C_2)/I_z+(C_1+C_2)/m\}D/U$$

$$+ \{L^2C_1C_2/U^2 - m(aC_1 - bC_2)\}D^0/(mI_z)$$

$$= 0 \qquad (3.4)$$

A convergent response to any disturbance will occur if the D^0 term is positive while a negative value shows a divergent response which indicates an unstable situation. The tyre characteristics are essentially positive, a is positive and b is positive, hence a negative value can only occur if, $aC_1 > bC_2$.

Note that the second term in the D^0 expression increases with forward speed, thus if the inequality given above is present there will be some speed below which the vehicle is stable and above which it is unstable.

$$U = L\surd[C_1C_2/\{m(aC_1 - bC_2)\}] \qquad (3.5)$$

This is the *critical speed* of the vehicle. Note that when $aC_1 < bC_2$ the calculation of equation (3.5) gives an imaginary answer which is the *characteristic speed*; the speed at which the understeer car is most sensitive to control.

3.5 TWO-DEGREE-OF-FREEDOM MODEL; DAMPING

The damping of the side-slip/yaw modes are given by the coefficient of the D^1 term in the characteristic equation

$$\{(a^2C_1 + b^2C_2)/I_z + (C_1 + C_2)/m\}/U \qquad (3.6)$$

The two parts $(a^2C_1 + b^2C_2)/I_z$ and $(C_1 + C_2)/m$ are approximately equal in value. Damping is inversely proportional to vehicle speed.

3.6 TWO-DEGREE-OF-FREEDOM MODEL; ROOTS OF THE CHARACTERISTIC EQUATION

For the oversteer car the roots are real and negative below the critical speed, and real and positive above this speed. The response time becomes increasingly important as speed increases.

The understeer car exhibits complex roots with damping inversely proportional to speed. The response time remains approximately constant.

3.7 TWO-DEGREE-OF-FREEDOM MODEL; STEADY STATE STEERING RESPONSES

It is convenient to use side-slip at the centre of gravity β (= $\arctan(V/U)$) instead of lateral velocity V. When $D(V)$ and $D(r)$ are zero the steady state equations are obtained. The following notation is used

$$Y_\beta = (C_1+C_2)\,;\quad N_\beta = (aC_1-bC_2+\hat{T}_1+\hat{T}_2)$$
$$Y_r = (aC_1-bC_2)/U\,;\quad N_r = (a^2C_1+b^2C_2+a\hat{T}_1-b\hat{T}_2)/U$$
$$Y_s = C_1\,;\quad N_s = aC_1+\hat{T}_1$$

The side-slip and yaw velocity responses may be obtained either in parametric form or in terms of the tyre characteristics and the mass and dimensions of the car.

$$\begin{bmatrix} Y_\beta & mU+Y_r \\ N_\beta & N_r \end{bmatrix}\begin{bmatrix} \beta \\ r \end{bmatrix} = \begin{bmatrix} Y_s \\ N_s \end{bmatrix} s \tag{3.7}$$

Side-slip and yawing velocity responses in parametric terms.

$$\beta/s\,|_{ss} = \frac{Y_sN_r - N_s(mU+Yr)}{Y_\beta N_r - N_\beta(mU+Y_r)}$$
$$r/s\,|_{ss} = \frac{Y_\beta N_s - N_\beta Y_s}{Y_\beta N_r - N_\beta(mU+Y_r)} \tag{3.8a}$$

In tyre and vehicle terms, with aligning moments neglected

$$\beta/s\,|_{ss} = \frac{bLC_1C_2-aC_1mU^2}{L^2C_1C_2-mU^2(aC_1-bC_2)}$$
$$r/s\,|_{ss} = \frac{ULC_1C_2}{L^2C_1C_2-mU^2(aC_1-bC_2)} \tag{3.8b}$$

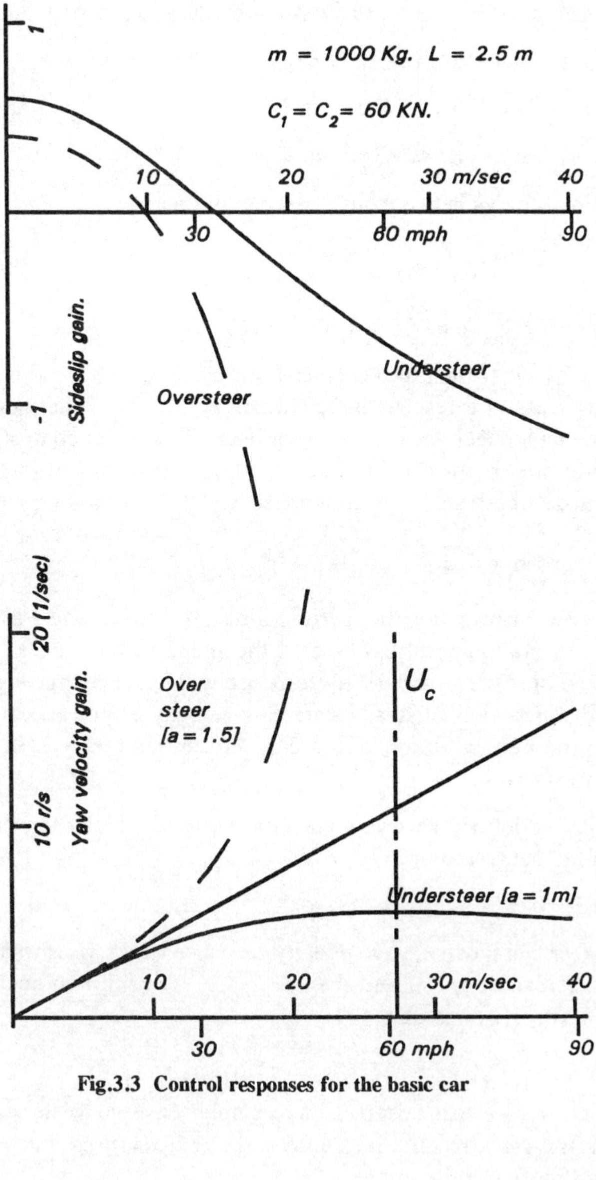

Fig.3.3 Control responses for the basic car

Another form of the responses is available by dividing throughout by LC_1C_2

$$\beta/s\,|_{ss} = -(b+AU^2)/(L+KU^2) \qquad (3.8c)$$

$$r/s\,|_{ss} = U/(L+KU^2)$$

$$A = (am/LC_2), \text{and } K = -m(aC_1-bC_2)/LC_1C_2$$

But the normal forces between the tyres and the road are

$$Z_1 = mb/L\ ; \quad Z_2 = ma/L$$

Hence

$$A = Z_2/C_2 \quad \text{and} \quad K = Z_1/C_1 - Z_2/C_2$$

When $aC_1 < bC_2$, then the characteristic speed is the speed at which the yaw velocity response reaches its maximum value, thus it defines the speed at which the understeer car is most sensitive to directional control.

The side-slip response is positive at low speeds but changes to a negative, nose in attitude, at high speeds. The cross over speed is given below.

$$U = \sqrt{\{lC_2/(am)\}} \qquad (3.9)$$

Figure 3.3 shows the theoretical side-slip angle and yaw velocity responses. In the hypothetical case of the speed of the car exceeding the critical speed then the yaw velocity response will again change sign.

Three distinct types of steady state yaw velocity response occur and may be termed the neutral steer, oversteer, and understeer yawing velocity responses to steering.

Neutral steer is defined as a yaw velocity response which increases linearly with speed, to achieve this

$$aC_1 = bC_2$$

An *understeer* car shows a yaw velocity response which rises to a maximum at the *characteristic speed* and then decreases. For this to occur then the equation (3.10.) requires that

$$aC_1 < bC_2$$

Finally, the *oversteer* car becomes increasingly sensitive to steering control, wind gusts and road irregularities until at the *critical speed* the yaw velocity response becomes infinitely large.

$$aC_1 > bC_2$$

3.8 THE CONSTANT RADIUS STEERING RESPONSE

A practical method of testing the responses of a vehicle involves driving it in a predictable manner with a driver. In this respect the steady state responses previously developed are not easily applied.

A realignment in the equations (3.7) allows the steady state curvature responses which represent the car being driven on a circular path at various speeds, to be calculated.

$$\begin{bmatrix} UY_\beta & -Y_s \\ UN_\beta & -N_s \end{bmatrix} \begin{bmatrix} \beta \\ s \end{bmatrix} = \begin{bmatrix} -(mU+Y_r) \\ -N_r \end{bmatrix} \begin{bmatrix} U/R \end{bmatrix} \qquad (3.10)$$

The equations (3.10) define the side-slip and steer angles as a car is driven at constant speed on a fixed radius.

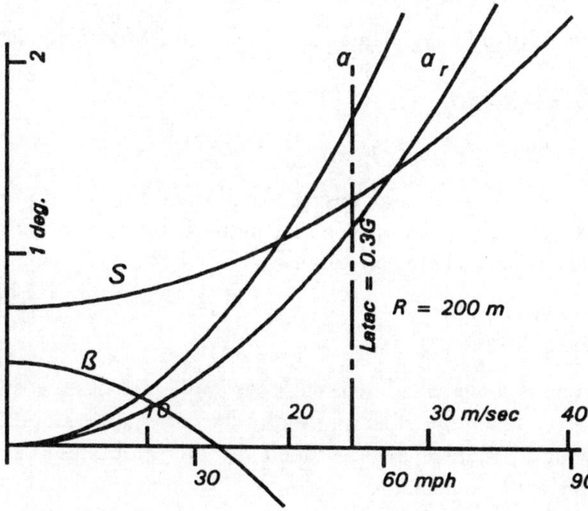

Fig.3.4(a) When the car is driven on a constant radius an increase in steer angle with speed implies understeer

Vehicle Data: m = 1000 Kg; I_z = 1650 Kg.m^2; L = 2.5 m; a = 1 m or 1.4 m
$dY/ds_f = dY/ds_r$ = 30 KN/rad $dN/ds_f = dN/ds_r$ = - 0.6KN.m/rad

The steering and side-slip responses for a vehicle driven at various speeds on a fixed radius turn are given first in terms of yaw and side-slip parameters and then as functions of tyre forces and vehicle dimensions.

$$\beta\,|_{ss} = \frac{-Y_r N_s + N_s (mU + Y_r)}{-Y_v N_s + N_v Y_s} (1/R)$$

$$s\,|_{ss} = \frac{-UY_v N_r + UN_v (mU + Y_r)}{-Y_v N_s + N_v Y_s} (1/R) \tag{3.11a}$$

In terms of the tyre and vehicle parameters

$$\beta\,|_{ss} = \frac{bLC_1C_2 - aC_1 mU^2}{LC_1C_2} (1/R)$$

$$s\,|_{ss} = \frac{L^2C_1C_2 - mU^2(aC_1 - bC_2)}{LC_1C_2} (1/R) \tag{3.11b}$$

Front and rear attitude angles may also be obtained for the constant radius test.

$$\alpha_f = s - \beta - a/R\,; \quad \alpha_r = -\beta + b/R$$

$$\alpha_f = bmU^2/(LC_1)\,; \quad \alpha_r = amU^2/(LC_2)$$

Both the side-slip and steer angles are defined for zero forward speed when the car is lined up on the circular path. These angles are frequently called the 'ideal' or 'Ackermann' angles.

Ideal steer angle: $s|_{U=0} = L/R$
Ideal side-slip angle: $\beta|_{U=0} = b/R$

The vehicle is *understeer* when the steer angle plot has a positive slope, and is *oversteer* when the steer angle plot has a negative slope. Note also that α_f is greater than α_r at any speed for the understeer case and the reverse is true for oversteer.

3.9 THE TWO-DEGREE-OF-FREEDOM CAR; TRANSIENT RESPONSES

The steady state responses show the final condition of a car due to the application of steering which moves to a given angle and is then held

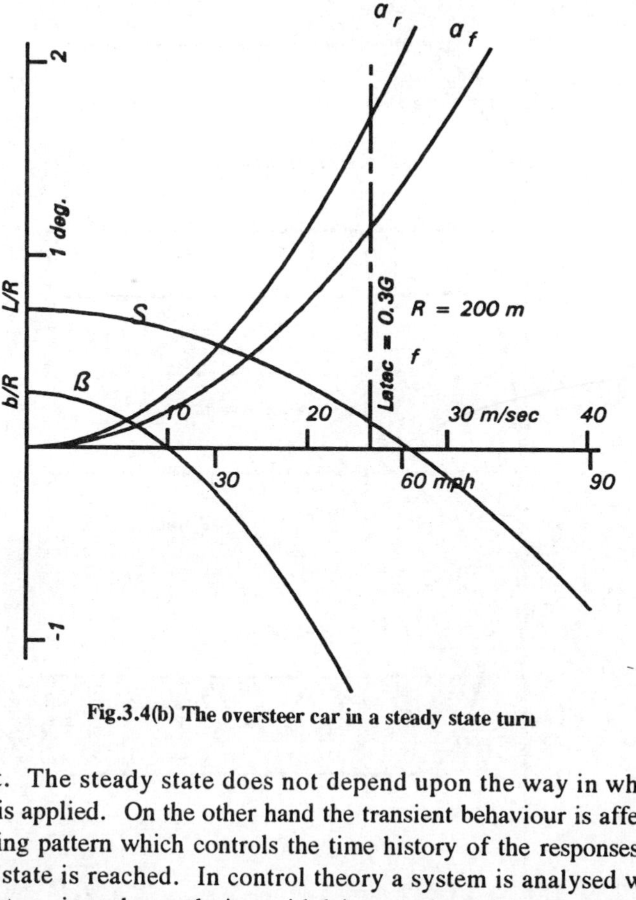

Fig.3.4(b) The oversteer car in a steady state turn

constant. The steady state does not depend upon the way in which the steering is applied. On the other hand the transient behaviour is affected by the steering pattern which controls the time history of the responses before a steady state is reached. In control theory a system is analysed with the typical step, impulse and sinusoidal input, the vehicle equations are examined in a similar fashion.

$$\dot{V} = [-(C_1+C_2)V/U - \{mU+(aC_1-bC_2)r/U\} + C_1 s]/m$$

$$\dot{r} = \{-(aC_1-bC_2+\hat{T}_1+\hat{T}_1)V/U - (a^2C_1+b^2C_2+a\hat{T}_1-b\hat{T}_1)r/U + aC_1 s\}/I_z$$

Response time as used in this text is the time for the system with exponential responses to rise to 1-1/e, 63.2 percent, of the steady state. This is also measured by the sub tangent to the curve at time t = 0. Some handling and control test procedures use 90 percent of the steady state value to define response time.

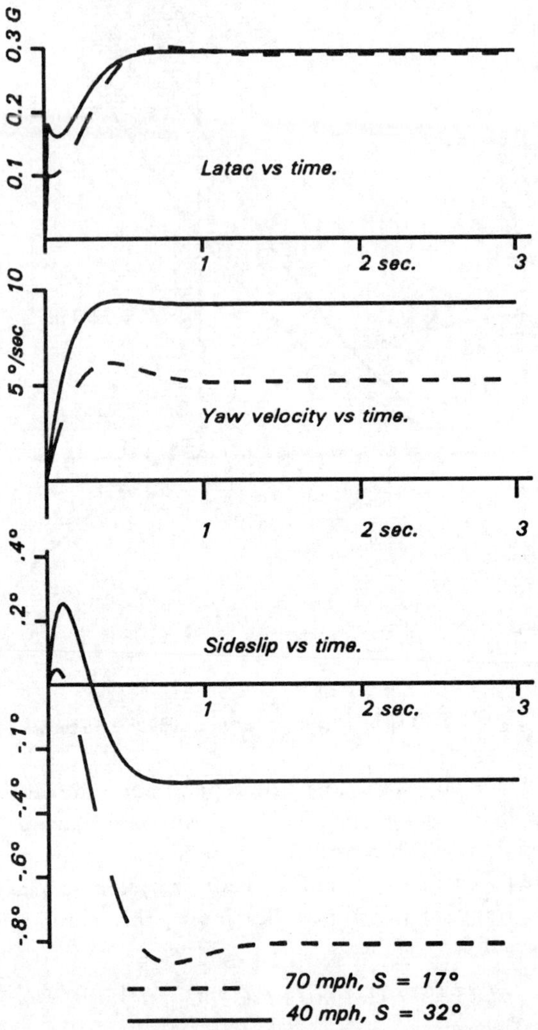

Fig.3.5(a) Handling responses; understeer car ($a = 1m$); characteristic speed = 61 mile/h. Vehicle data as before

In this particular example a steering ratio of 17 has been taken. The steering system is assumed to be extremely stiff and without inertia. Later the effects of including the steering as a dynamic sub-system will be discussed.

The equation of motion are examined by using time histories of side-slip, yaw velocity, etc. These are produced by digital simulation. Lateral acceleration of the origin for the vehicle axes is given in these results because although it is not a variable it is an easily measured value in practical tests and can be related to the basic variables.

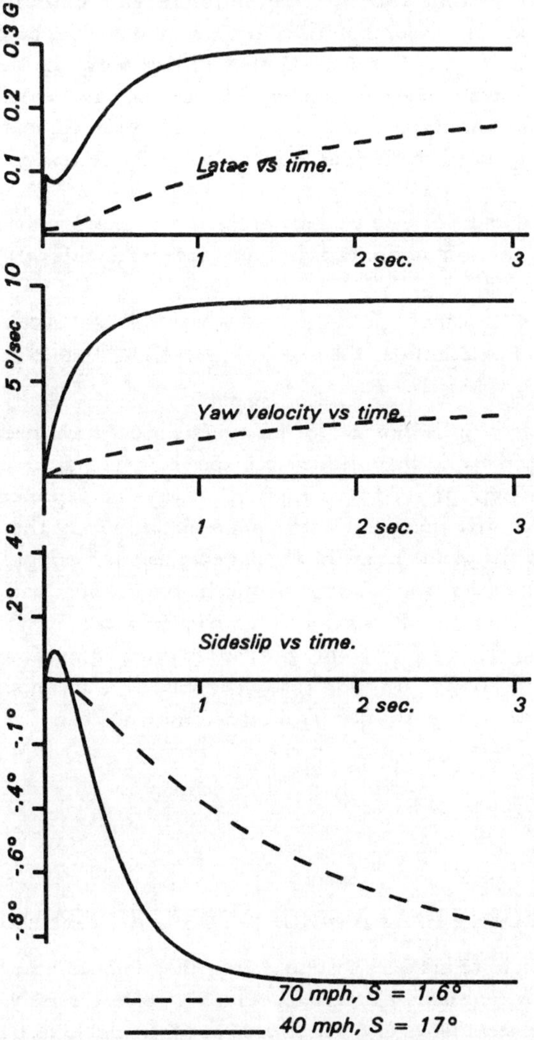

Fig.3.5(b) The oversteer car (a = 1.4m); critical speed = 79 mile/h

Alternative methods are Nyquist, or Bode plots and by calculating the roots of the equations.

Understeer car

At the moderate speed of 40 mile/h the steady state conditions are reached in less than 1 second and the overshoot in yaw velocity is barely discernable. Lateral acceleration rises to a steady state with no overshoot.

The high speed test, 70 mile/h, applies 17 degrees at the steering wheel to give a 0.3 G steady state. The overshoot on the yaw velocity and side-slip time histories are typical and show the decrease in damping associated with increasing speed. Note that the steady states are reached in about 1 second.

The reduced value of yaw velocity at the higher speed is the direct result of basing the test on a constant level of lateral acceleration. Lateral acceleration = Ur.

The ratio of steering angle at the road wheel to lateral acceleration is 6.3 degrees/G at 40 mile/h; this falls to 3.3 degrees/G at 70 mile/h.

Oversteer car

Steady state steering responses for an oversteer car have been shown, in Fig.3.5, to increase in magnitude with speed. The transient part of the response to a step input of steering will always be exponential but the response time will increase with speed as shown by the diagram of Fig.3.6(b). This results in a car which becomes increasingly sensitive to steering while requiring a longer time to reach the steady state (3.3 degrees/G, $t_e < 0.5$ s at 40 mile/h; 3 degrees/G, $t_e > 1.5$ s at 70 mile/h). The oversteering car becomes more difficult to drive at higher speeds due to these responses. Wind gusts and road disturbances result in similar effects to steering, so a car is not generally designed for oversteer.

3.10 SIDE-SLIP AS A SINGLE-DEGREE-OF-FREEDOM

it is interesting to reduce the system still further and examine the side-slip and yawing angle responses separately with a view to precisely defining the role of tyre characteristics and wheelbase. For the vehicle restrained in yaw the significance of the lateral tyre is highlighted.

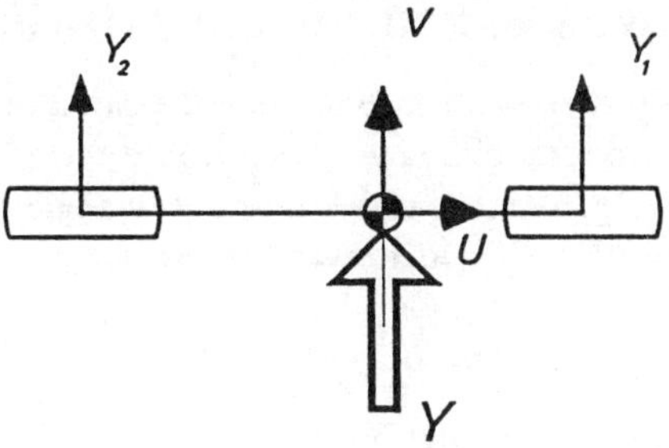

a) Sideslip mode

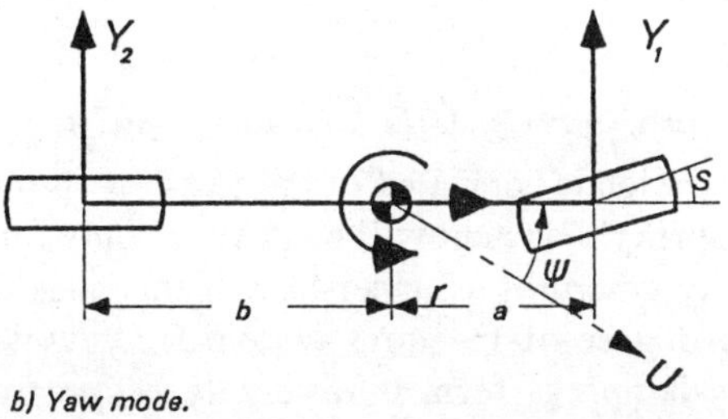

b) Yaw mode.

Fig.3.6 Side-slip and yaw as single degrees of freedom

$$mV = (C_1+C_2)V/U + Y \tag{3.12}$$

The solution of this equation shows the vehicle moving sideways on the road as the result of an applied force *Y*. The lateral velocity tends exponentially to a steady value.

$$\beta = A(1-e^{(-pt)}) \tag{3.13}$$

Where

$$A = -Y/(C_1+C_2)$$

$$p = -(C_1+C_2)/mU$$

Thus, under the action of a lateral force the vehicle will accelerate sideways until the side-slip velocity reaches a value such that the slip angle of the vehicle causes the tyre forces to balance the applied force. This small demonstration illustrates the importance of tyre cornering stiffness in maintaining vehicle path within a traffic lane. In a practical situation the car is yawed at the body slip angle so as to follow a given path in the presence of a side wind or cambered road.

3.11 YAW AS A SINGLE-DEGREE-OF-FREEDOM

Restraint in side-slip is equivalent to pinning the car at the origin of the axis set so that it is free to rotate in yaw. At each tyre the attitude angle will be composed of a steering component due to the yaw angle together with a slip angle from the yaw velocity and forward speed vectors. A sketch is given in Fig.3.7.

$$I_z r = aC_1(s+\psi-ar/U) - bC_2(\psi+br/U) \tag{3.14}$$

Let the car have no initial angular heading, then

$$\psi = \int r dt$$

Thus

$$[I_z D^2+\{(a^2C_1+b^2C_2)/U\}D-(aC_1-bC_2)]=aC_1 s \tag{3.15}$$

When this equation is compared to the classical spring-mass system it is clear that the 'spring' constant will be positive only for the condition $aC_1 < bC_2$, previously described as understeer. If this condition is not satisfied then the single-degree-of-freedom system for yaw will be divergently unstable. The 'damping' term is inversely proportional to speed and proportional to the square of the wheelbase.

These very basic single-degree-of-freedom models show clearly that the lateral movement on the road will always be exponential and that the rate of lateral movment is dependent upon the ratio of lateral tyre stiffness to mass. The yawing motion is exponential at low speed and becomes sinusoidal as the damping decreases. The equation stresses the importance of a long wheelbase to dampen the yaw motion. Thus the well tempered car will have a high ratio $(C_1+C_2)/m$ to ensure good side-slip response and a value L^2/I_z, depending upon the decision of the designer to have a quick, or more leisurely response to steering.

3.12 NEUTRAL STEER POINT

This term used to describe the position along the vehicle axis at which an applied lateral force produces side-slip without a yawing motion, this distance l' will be determined from the front axle of the vehicle. Since the tyre forces which result are generated with equal slip angles at front and rear then

$$l'C_1 = (L-l')C_2$$

or

$$l' = C_2 L/(C_1+C_2) \tag{3.16}$$

3.13 STATIC MARGIN

This is the distance between the neutral steer point and the centre of gravity of the vehicle.

$$SM = a - l'$$

Hence

$$SM = (aC_1 - bC_2)/(C_1 + C_2) \tag{3.17}$$

Thus the static margin is another measure of the understeer/oversteer of a car.

3.14 THE REAR STEERED CAR

Rear wheel steering is frequently employed in construction vehicles. In road vehicles the rear wheel steer effect is usually confined to kinematic changes produced by roll angle and compliance. The equations of motion are similar except that the steering term appears in the rear attitude angle.

$$\begin{aligned} m(\dot{V} + Ur) &= C_1 \{ - (V+ar)/U\} + C_2 \{s_r - (V-br)/U\} \\ I_z\dot{r} &= (T_1 + aC_1)\{ - (V+ar)/U\} + (T_2 - bC_2)\{s_r - (V-br)/U\} \end{aligned} \tag{3.18}$$

Note that the characteristic equation is not changed and therefore the previous statements on stability still apply. It is clear from consideration of static equilibrium that the lateral forces from the tyres in a given turn are not altered although the components of the attitude angles of the tyres will be distributed differently. Thus while in the steady state turn the attitude angles are similar to those of the steered front wheel car the steer angle and side-slip velocity will be different.

The steady state steering and side-slip required to keep the vehicle on a specified radius are given below.

$$\begin{aligned} s &= \frac{mU^2(aC_1 - bC_2) - L^2C_1C_2}{LC_1C_2} \; (1/R) \\ \beta &= \frac{-(mU^2bC_2 + aLC_1C_2)}{LC_1C_2} \; (1/R) \end{aligned} \tag{3.19}$$

These equations show that the steering is reversed and the side-slip angle is changed, starting as a negative value at zero speed, ($-a/L$), and becoming more negative as speed increases. The basic definition of understeer for the front steer vehicle is that the steer angle required for a fixed radius turn shall increase with speed thus bC_2 is larger than aC_1. With the rear steer configuration the same vehicle will now exhibit an increasingly negative steer angle in the constant radius test.

3.15 ALL-WHEEL STEERING

The equations for a car with steering at both front and rear wheels are simply a combination of the two separate cases.

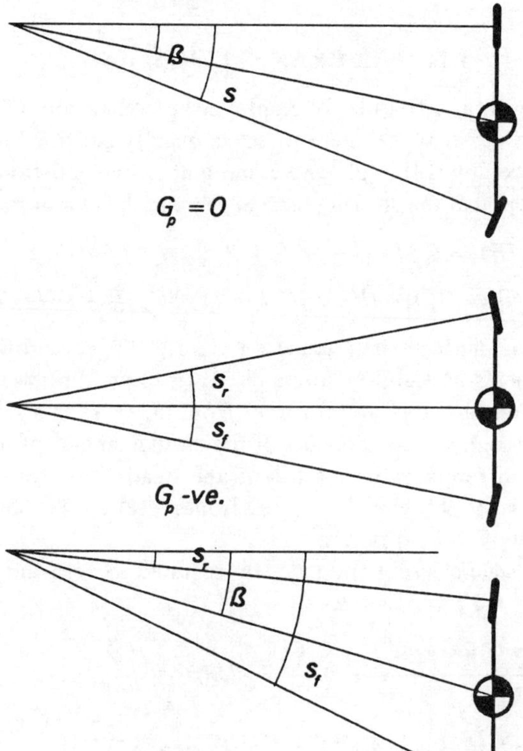

Fig.3.7 The ideal, or Ackermann angles for various four wheel steering ratios with the vehicle at slow speed on a curve

The 'ideal' or Ackermann angles required for kinematic rolling of the tyres on a given radius have been defined. When the rear wheels are also steered these angles are changed. Figure 3.7 shows the Ackermann angles for some combinations of front and rear steering. Note that position of the zero speed centre of turn shifts, moving forward when the rear wheels steer in opposite rotation to the front wheels and rearward when the steering rotations are in the same sense. The vehicle side-slip angle is also affected.

$$\begin{aligned} m(\dot{V}+Ur) &= C_1\{s_f-(V+ar)/U\} + C_2\{s_r-(V-br)/U\} \\ I_z\dot{r} &= (\hat{T}_1+aC_1)\{s_f-(V+ar)/U\} - (\hat{T}_2-bC_2)\{s_r-(V-br)/U\} \end{aligned} \tag{3.20}$$

Again the lateral tyre forces required for equilibrium are not changed by the steering system, thus the steering ratios at front and rear may be manipulated to change the side-slip angle of the body in the steady state and to produce any desired transient. Let the ratio of rear to front steering be G_p, then the steady state curvature responses are

$$\begin{aligned} s &= \frac{mU^2(aC_1-bC_2)- L^2C_1C_2}{(G_p-1)LC_1C_2} \quad (1/R) \\ \beta &= \frac{mU^2(aC_1-bG_pC_2)-LC_1C_2(b+aG_p)}{(G_p-1)LC_1C_2} \quad (1/R) \end{aligned} \tag{3.21}$$

It is possible to develop a steering ratio G_p so that the steady state side-slip is zero at all speeds, assuming linear tyre coefficients.

$$G_p = \frac{(amU^2/LC_2) - b}{(bmU^2/LC_1) + a} \tag{3.22}$$

Since am/l is the vehicle mass at the rear axle (m_2) and bm/L is the mass at the front axle (m_1) then

$$G_p = \frac{(m_2/C_2)U^2 - b}{(m_1/C_1)U^2 + a} \tag{3.23}$$

From inspection of Fig.3.7 it would appear that a negative value for G_p may be useful for low speed movement in restricted space. Figure 3.8 simply shows that the understeer vehicle in a steady state turn has a degree of understeer which is modified by four wheel steering (4WS).

A transient model is used to demonstrate some of the changes to be

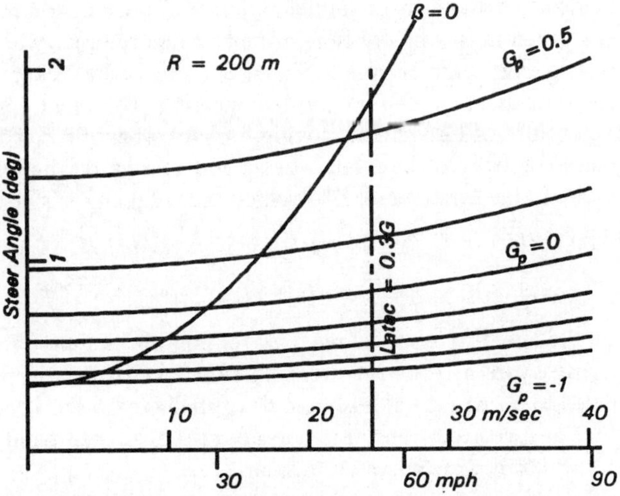

Fig.3.8 Steer angle vs speed for a range of G_p. The steer angle to maintain ß = 0 is also shown.

m = 1000 kg; $C_1 = C_2$ = 60 kN/rad; L = 2.5m; a = 1.2m

expected from four wheel steer vehicles. The results are presented as time histories of lateral acceleration, yaw velocity and side-slip. The amplitude of the single sinusoid steering input is selected to give 0.3G lateral acceleration in a steady state at 70 mile/h. In all cases the time delays for the first and second peak values are similar. Negative rear steer gives lower values of the second peak for all the results. This may indicate an improvement in the use of tyre forces in this phase of a reverse steering test.

3.16 STEERING SYSTEM DYNAMICS

The flexibility of the steering mechanism and the inertia of the road wheels oscillating around the steering ball joints can produce effects which considerably change the driver's perception of the car. By increasing the amount of steering rotation for the steering handwheel in a steady state turn due to linkage flexibility an understeering feel is produced while the transient characteristics also change particularly in sinusoidal and random steer tests. It will be assumed that the steering handwheel is under the control of the driver at all times and the handwheel is not part of the dynamic system. This provision that the steering wheel is firmly held allows calculation of steering torque and position.

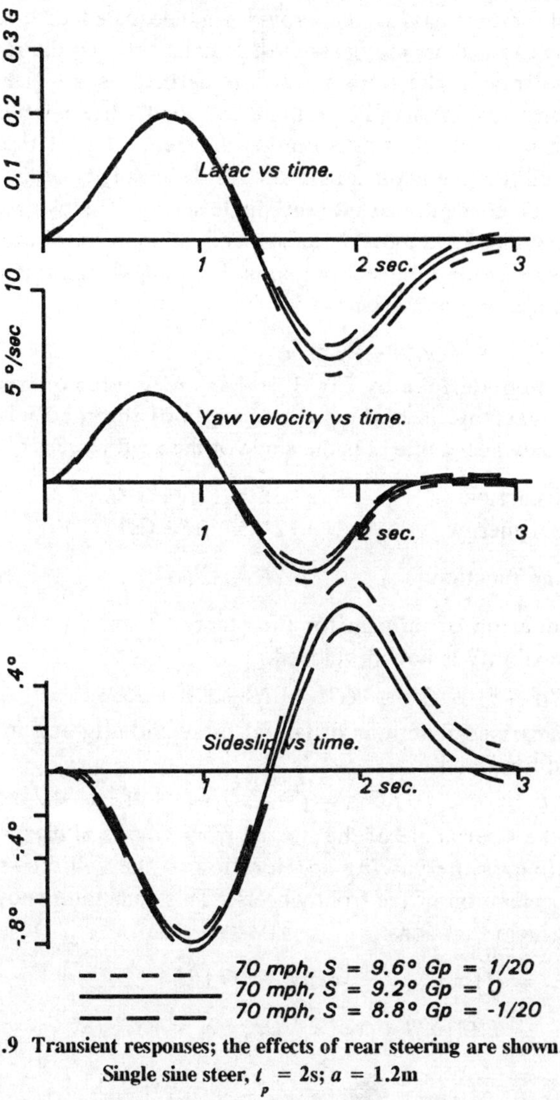

Fig.3.9 Transient responses; the effects of rear steering are shown. Single sine steer, t_p = 2s; a = 1.2m

I' = inertia of the road wheels and drag link about the steering joints.
K = steering system stiffness.
C' = steering system damping.
G = steering box ratio (handwheel/road wheel).

A single spring-mass system provides an adequate model for the steering gear in this case; damping is assumed to exist between the road wheels and the vehicle frame. The steering ratio is defined as the ratio between the angles turned by the handwheel and the road wheels when no external forces are applied, thus G includes the geometry of the linkage. An efficiency of 100 percent is assumed. All the elasticity of the mechanism is assumed to be concentrated into the single spring. The two road wheels and the connecting linkage move in unison and the moment of inertia, I', is that of these parts around the steering axis. Lagrange's method is employed to derive the equation of motion.

$$\partial T/\partial \dot{q} - \partial V/\partial q - \partial F/\partial \dot{q} = Q$$

The system defined by Fig.3.10 has one degree of freedom and the particular variable is s, the steered angle of the road wheels. S is the steering handwheel angle, s is the angle of the road wheels.

Kinetic energy: $2T = I'(r + \dot{s})^2$

Potential energy: $2V = K(S - Gs)^2$

Damping function: $2F = C'(\dot{s})^2$

The equation of motion for the steered front wheels of a car with steering flexibility is now developed.

$$I'(\dot{r} + \ddot{s}) + C'\dot{s} + KG^2s = N + x'Y + KGS$$

The linearised functions of lateral force and aligning moment may be substituted for Y and N.

$$I'(\dot{r}+\ddot{s})+C'\dot{s}+KG^2s = (\hat{T}+Cx')\{s-(V+ar+x's)/U\}+KGS \quad (3.24)$$

Since the steer angle of the steered front wheels is measured relative to the vehicle axis the yawing acceleration of the vehicle is added to the angular acceleration of the front wheels. This equation is now incorporated into the vehicle equations.

$$S_v = \hat{T}_1+x'C_1; \quad S_r = a(\hat{T}_1+x'C_1)/U$$

$$S_{ds} = [\{C+x'\hat{T}_1+(x')^2C_1\}/U]; \quad S_s = KG^2-\hat{T}_1-x'C_1$$

$$\begin{bmatrix} mD+Y_v & mU+Y_r & -Y_s \\ N_v & I_zD+N_r & -N_s \\ S_v & I'D+S_r & I'D^2+S_{ds}D+S_s \end{bmatrix} \begin{bmatrix} v \\ r \\ s \end{bmatrix} = \begin{bmatrix} 0 \\ 0 \\ KG \end{bmatrix} \begin{bmatrix} S \end{bmatrix} \quad (3.25)$$

In the steady state all the derivative terms are zero. The curvature responses are given by the following equation.

$$\begin{bmatrix} UY_v & -Y_s & 0 \\ UN_v & -N_s & 0 \\ US_v & S_s & -KG \end{bmatrix} \begin{bmatrix} \beta \\ s \\ S \end{bmatrix} = \begin{bmatrix} -(mU+Y_r) \\ -N_r \\ -S_r \end{bmatrix} \begin{bmatrix} U/R \end{bmatrix}$$

Thus the responses are

$$\beta = \frac{\{-Y_r N_s + N_s(mU+Y_r)\}}{(-Y_v N_s + N_v Y_s)} (1/R)$$

$$s = \frac{\{-UY_v N_r + UN_v(mU+Y_r)\}}{(-Y_v N_s + N_v Y_s)} (1/R) \tag{3.26}$$

$$S = \frac{Y_v(N_s S_r - N_r S_{ds}) - N_v\{Y_s S_r - S_{ds}(mU-Y_r)\} + S_v\{Y_s N_r - N_s(mU-Y_r)\}}{KG(-Y_v N_s + N_v Y_s)} (1/R)$$

The side-slip and steer angles of the front wheels are similar to those obtained for the linear two-degree-of-freedom car while the angle of the steering handwheel is greater than before due to the flexibility of the steering system. Thus when the flexure of the steering system is considered the effect at the handwheel will be to increase the apparent understeer.

Steering handwheel torque is obtained by consideration of the work done.

Work in = Work out

$$TdS = (x'Y+N)ds$$

But

$$dS/ds = G$$

Therefore

$$T = (x'Y+N)/G \tag{3.27}$$

In the transient mode the effect of the steering flexibility is to generate a time lag in the application of steer angle at the front wheels as well as modifying the actual angle.

3.17 THE FREE CONTROL CAR

Not only do the steered road wheels possess a degree of freedom but the steering wheel inertia also plays a part in the control of a vehicle. This analysis demonstrates that contribution. The steering wheel is now part of a two-degree-of-freedom system. The analysis is an extension of the previous case, the steering handwheel angle, S, is now the second variable and an allowance is made for damping in the bearing which supports the steering column beneath the steering wheel.

Kinetic energy: $2T = I_{hw}\dot{S}^2 + I'(r + \dot{s})^2$

Potential energy: $2V = K(S - Gs)^2$

Damping function: $2F = C_{hw}(\dot{S})^2 + C'(\dot{s})^2$

The equations of motion are now obtained from the repeated use of Lagrange's equation in the form shown previously.

$$I_{hw}\ddot{S} + C_{hw}\dot{S} + KS - KGs = 0 \qquad (3.28)$$

$$I'(\dot{r}+\ddot{s})+C'\dot{s}+KG^2s-KGS = (\hat{T}+x'C)\{s-(V+ar+x'\dot{s})/U\}$$

Although the car cannot be directed without some restraint on the steering wheel the effect of the driver is somewhat like a spring attached to the rim of the wheel and with displacement applied at the other end of the spring. The free oscillations of the vehicle and steering mechanism are interesting since it is quite possible to make a car unstable at low speeds (40 mile/h) by increasing the inertia of the steering wheel, thus the practice of mounting a second steering wheel onto the original unit to measure torque and angle changes the characteristics of the car and is not recommended as a test method. Friction and damping play a significant part in controlling these oscillations.

$$\begin{bmatrix} mD+Y_v & mU+Y_r & -Y_s & 0 \\ N_v & I_zD+N_r & -N_s & 0 \\ S_v & I'D+S_r & I'D^2+S_{ds}D+S_s & KG \\ 0 & 0 & -KG & I_hD^2+C_hD+K \end{bmatrix} \begin{bmatrix} v \\ r \\ s \\ S \end{bmatrix} = \begin{bmatrix} 0 \end{bmatrix} \qquad (3.29)$$

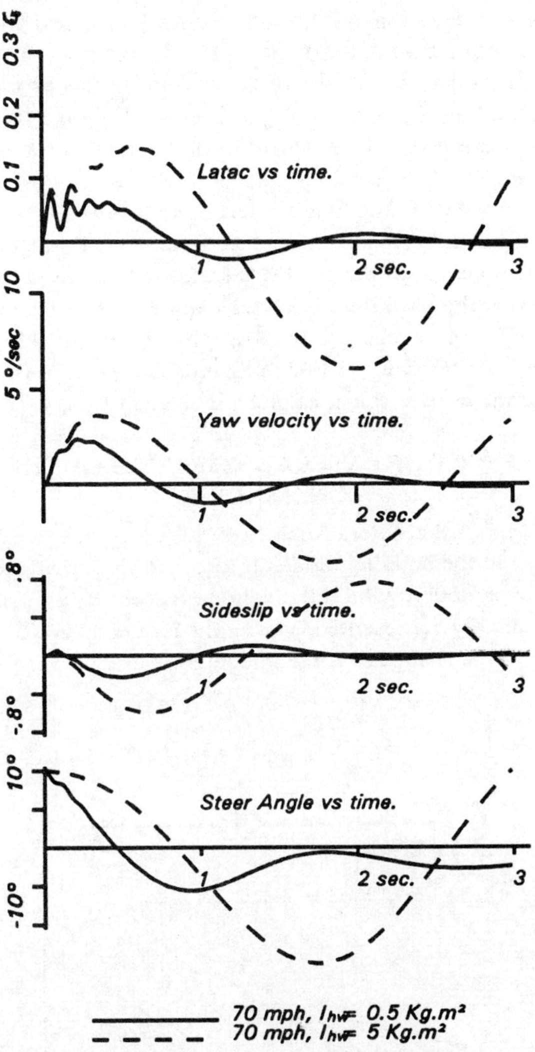

**Fig.3.10 When the steering handwheel is rotated and then released the subsequent oscillation may decay or expand.
Speed and handwheel inertia are important factors**

With a set of equations of this size the only practical way to study the stability is to obtain numerical values for the roots of the characteristic equation or to set up a simulation of the system. In this case a simulation is used to demonstrate some possible problems. Figure 3.10 is based on a stable, understeering vehicle in which the only potential for instability is the steering system.

With a low value of handwheel inertia the oscillation of the vehicle all decay. The initial high frequency content of the lateral acceleration curve is due to the introduction of the handwheel angle as an initial condition of the simulation with the road wheels at zero angle. When the test is repeated with the inertia increased the oscillations show the vehicle becoming uncontrollable. Damping at the road and handwheels is maintained at 50 percent of the critical value, as given in the equations (3.28)

3.18 EFFECT OF LATERAL STIFFNESS OF THE TYRES ON HANDLING

It has been shown that lateral distortion of a tyre due to cornering force causes a delay in the build up of lateral force after a steering disturbance.

For many models in which the vehicle travels at high speed the delay may be adequately represented by simply reducing F&M by an amount proportional to the change over the integration period (see section 1.22)

$$FY = FY - (FY - FY_{LAST}) \times EXP(- DELAY)$$

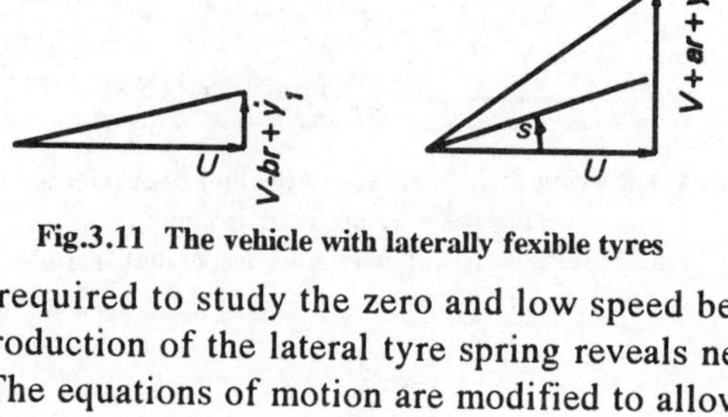

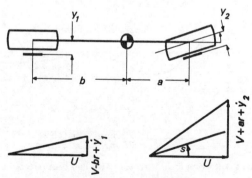

Fig.3.11 The vehicle with laterally fexible tyres

When it is required to study the zero and low speed behaviour of a vehicle the introduction of the lateral tyre spring reveals new aspects of performance. The equations of motion are modified to allow for the fact that there are now springs between the tyre/road contact surfaces and the vehicle. The car now has two modes of vibration while it is stationary; a lateral shake and a torsional oscillation around the polar axis. The apparent damping inceases with speed and the car responses become exponential at higher speeds.

$$m(\dot{V}+Ur) = K_1 y_1 + K_2 y_2$$

$$I_z \dot{r} = aK_1 y_1 - bK_2 y_2$$

$$K_1 y_1 = C_1(s - \{V+ar+\dot{y}_1\}/U)$$

$$K_2 y_2 = C_2\{-(V-br+\dot{y}_2)/U\}$$

A solution which uses the Routh - Hurwicz stability criteria is demonstrated with this vehicle model. Let $\lambda_1 = UK_1/C_1$ and $\lambda_2 = UK_2/C_2$

$$\begin{bmatrix} mD & mU & -K_1 & -K_2 \\ 0 & I_z D & -aK_1 & bK_2 \\ 1 & a & D+\lambda_1 & 0 \\ 1 & -b & 0 & D+\lambda_2 \end{bmatrix} \begin{bmatrix} V \\ r \\ y \\ y \end{bmatrix} = \begin{bmatrix} 0 \\ 0 \\ U \\ 0 \end{bmatrix} s \qquad (3.30)$$

The characteristic polynomial of the system is a quartic.

$$A_4 D^4 + A_3 D^3 + A_2 D^2 + A_1 D + A_0 = 0$$

where

$$A_4 = mI_z$$

$$A_3 = mI_z(\lambda_1 + \lambda_2)$$

$$A_2 = mI_z\lambda_1\lambda_2 + m(a^2K_1 + b^2K_2) + I_z(K_1 + K_2)$$

$$A_1 = m(a^2K_1\lambda_2 + b^2K_2\lambda_1) - mU(aK_1 - bK_2) + I_z(K_1\lambda_2 + K_2\lambda_1)$$

$$A_0 = -m(aK_1\lambda_2 - bK_2\lambda_1) + L^2K_1K_2 \qquad (3.31)$$

Since the term *UK/C* is similar for front and rear tyres the coefficients may be reduced, and an approximation made that the quartic is bi-quadratic for the case that the centre of gravity is at the centre of the wheelbase.

$$(D^2 + \lambda D + KL^2/2I_z) \cdot (D^2 + \lambda D + 2K/m) = 0 \qquad (3.32)$$

Equation (3.32) shows two motions, a lateral shake and a torsional oscillation, both independent of speed. The damping in each mode increases with speed. Thus the car has two undamped motions when stationary and both modes will become exponential as the speed increases.

The steady state responses to steering now include terms for the lateral tyre deflections, the yaw velocity response may be shown to be similar to that for the rigid-tyred vehicle.

$$\text{Side-slip response:} \qquad r\beta/s\,\big|_{ss} = \frac{bLK_1K_2 - mUaK_1}{L^2K_1K_2 - mU\mu(aK_1-bK_2)}$$

$$\text{Yaw velocity response:} \qquad r/s\,\big|_{ss} = \frac{ULK_1K_2}{L^2K_1K_2 - mU\mu(aK_1-bK_2)} \tag{3.33}$$

The curvature responses for a fixed radius of turn are obtained from the matrix given below.

$$\begin{bmatrix} 0 & 0 & -K_1 & -K_2 \\ 0 & 0 & -aK_1 & bK_2 \\ 1 & -U & \lambda & 0 \\ 1 & 0 & 0 & \lambda \end{bmatrix} \begin{bmatrix} v \\ s \\ y_1 \\ y_2 \end{bmatrix} = \begin{bmatrix} -mU^2 \\ 0 \\ -aU \\ bU \end{bmatrix} I/R$$

3.19 CONSTANT RADIUS STEERING RESPONSE WITH NON-LINEAR TYRE CHARACTERISTICS

When a car is driven on a constant radius turn the demand for lateral force from the tyres increases rapidly with speed and the tyre F&M curves enter the non-linear regime. Under these conditions it is usual to examine the force and moment balance for the vehicle and deduce the tyre attitude angles by cross referencing the tyre F&M curves.

Note that the car is shown in a position which is not tangential to the curve. The angle ß in the sketch has the car pointing into the curve, typical of a high lateral force condition when the side-slip angle is negative. This side-slip angle is formed by the addition of the lateral velocity and forward speed vectors. At very low speeds where inertial forces are negligible the rear axle lies on the radius *R* and the steered angle of the front wheels is

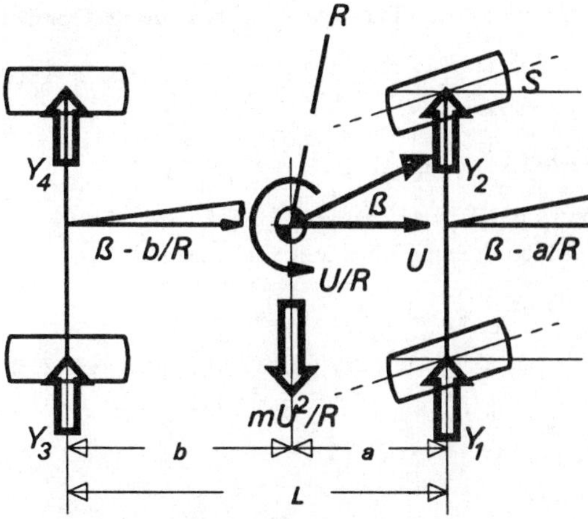

Fig.3.12 Force and moment (F&M) equilibrium in a steady state turn

arranged so that the tangents to the paths of these wheels intercepts the radial line at the centre of the turning circle. This is the 'ideal' or Ackermann position to minimize tyre scrub at low speeds on sharp turns. For most conditions the obliquity of the lateral acceleration vector can be ignored, then the following equations apply

$$mU^2/R = Y_1 + Y_2 + Y_3 + Y_4 \tag{3.34}$$

$$0 = a(Y_1+Y_2) - b(Y_3+Y_4) + N_1+N_2+N_3+N_4$$

The effects of the aligning moments N_1 and N_2 which are due to small rearward offsets of the lateral forces from the centers of the road/tyre interfaces, will be included in the values of a and b in equations (3.12) by adding the term x_p to a and b. Thus

$$Y_f = [mU^2/R](b/L)$$

$$Y_r = [mU^2/R](a/L) \tag{3.35}$$

where

$$L = a + b$$

Equation (3.34) gives the lateral forces required to hold a vehicle in a steady state turn of specified radius and speed. Once these forces have been determined the attitude angles necessary to develop them are obtained from the tyre characteristic curves. The front and rear attitude angles have been defined previously.

$$\alpha_f = s - (V+ar)/U$$

$$\alpha_r = - (V-br)/U$$

The yawing velocity for a vehicle at known speed on a given radius is simply $r = U/R$; hence the side-slip angle is obtained.

$$\beta = -\alpha_r + b/R \qquad (3.35a)$$

Moving to the front axle; the only unknown is now the steer angle.

$$s = \alpha_f + B + a/R \qquad (3.35b)$$

or

$$s = \alpha_f - \alpha_r + (a+b)/R \qquad (3.35c)$$

tyre characteristics are assumed to be in the cubic form given in equation (1.57) which is repeated here.

$$Y = Cs - C^2 s|s|/(3Ym) + (Cs)^3/(3(3Ym)^2)$$

This tyre data is measured at one value of normal force Z_n when the initial slope of the lateral force vs steer angle curve is $C=dY/ds$. Force and moment may be adjusted to account for different normal force, friction, and fore/aft force using formulae given below.

Normal force ratio:

$$\Gamma = Z/Z_n$$

Maximum lateral force is $f(\mu,\Gamma)$:

$$Ym = 3\mu Z/2(\Gamma - \Gamma^2/3)$$

Lateral force is decreased by fore/aft force:

$$Ym = (Ym^2 - X^2)$$

Initial slope is $g(\Gamma)$:

$$C = 2/3*dY/ds(1 + \Gamma - \Gamma^2/2)$$

Steer angle for maximum lateral force:

$$s_m = 3Ym/C$$

Pneumatic trail decreases with angle:

$$x_p = (dN/ds)/(dY/ds)(1 - |s/s_m|) \qquad (3.36)$$

These formulae have been shown in a form suitable for use in a computer programme in section 1.22.

The measured value of *Ym* is related to the normal force by a friction number, the maximum lateral force can be modified by introducing a different friction number. Normal force for a tyre is calculated from the combination of loads due to weight distribution, load tranfer due to lateral acceleration, longitudinal forces, and aerodynamic forces. When the two wheels on an axle are considered the effect of lateral acceleration load transfer is to reduce the lateral force available at a given attitude angle. This is shown in Fig.3.13.

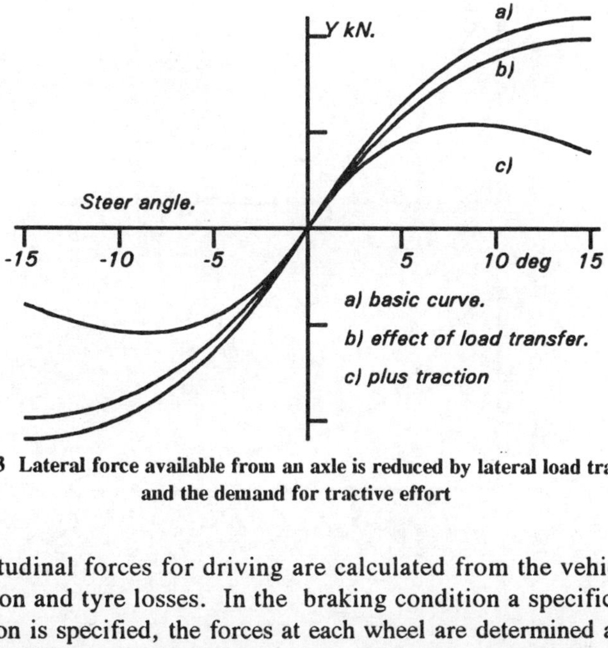

Fig.3.13 Lateral force available from an axle is reduced by lateral load transfer and the demand for tractive effort

Longitudinal forces for driving are calculated from the vehicle drag, acceleration and tyre losses. In the braking condition a specific level of deceleration is specified, the forces at each wheel are determined according to the brake design and the limiting values. The value of *Ym* is then adjusted on the assumption that the force in the road plane not required for braking is available for cornering; the 'friction ellipse' concept.

Small perturbations about any non-zero lateral acceleration are defined by the equations given previously, however, the tyre coefficients are now the local values at the particular Lateral acceleration. Thus, it is possible to define the critical/characteristic speed, static margin, and other measures at any vehicle condition.

3.20 SINUSOIDAL STEERING RESPONSES

Sinusoidal excitation is a standard test input for a dynamic system. In the development of vehicle handling tests the proposal has been made to excite the vehicle by steering in a sinusoidal manner. Several schemes have been considered, single sine steering, continuous steering at a single frequency and amplitude, random steering, and steering by 'sweeping' a range of frequencies. The lane change may also be considered as a form of sine steer.

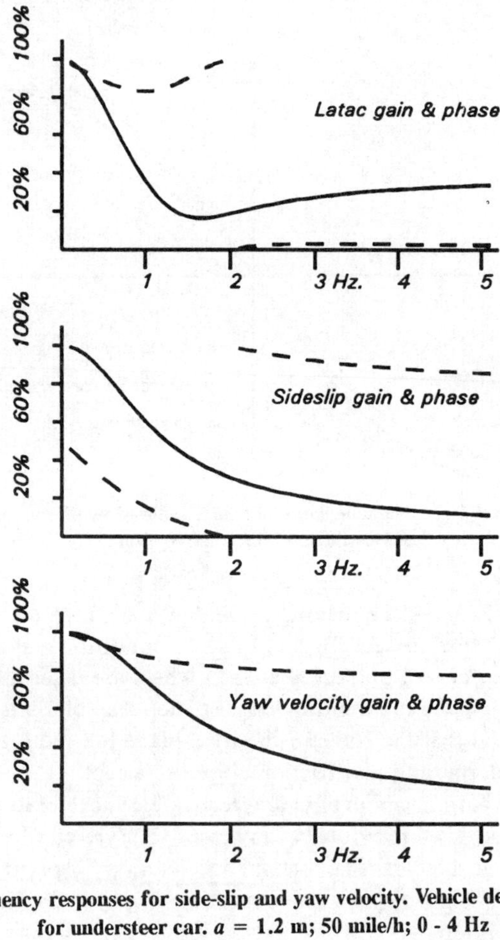

Fig 3.14 Frequency responses for side-slip and yaw velocity. Vehicle details as previously for understeer car. a = 1.2 m; 50 mile/h; 0 - 4 Hz

The single sine/lane change steering pattern at a specified frequency is illustrated by the simulation of the effects of steering flexibility. Interest is centered on the delays in reaching the first and second peaks and in the relative amplitude of these peaks.

Random steering and sweeping through a range of frquencies may be considered as variations of the same procedure. In both cases the practical difficulty of obtaining sufficient energy in the very low frequency end of the test exists and for this reason the process of testing at an individual steering frequency may be preferred.

Figure 3.14 shows the results of solving the basic equations of motion for sinusoid steering.

$$m(\dot{V}+Ur) = C_1\ \{S.sin(pt)-(V+ar)/U\} + C_2\{-(V-br)/U\}$$

$$I_z\dot{r} = (aC_1+\hat{T})\{S.sin(pt)-(V+ar)/U\} - (bC_2+\hat{T})\{-(V-br)/U\}$$

These solutions are obtained by assuming an amplitude and phase for both side-slip and yaw velocity and their derivatives. Details of the method may be found in texts on control engineering.

The diagrams of Fig.3.14. are drawn as percentage of the steady state value. Phase as shown is divided by 360, thus when the dashed line is in the upper part of the graph there is a phase lag. Lateral acceleration gain decreases to a minimum around 1.8 Hz and then rises slowly with a phase advance of some 10 degrees.

The yaw velocity decreases with steer frequency, starting in phase with steering and tending to 90 degrees lag as frequency increases.

Side-slip, ß, starts at 180 degrees phase angle which then falls to 0 degrees around 2 Hz and finally tends to lag as frequency increases further. The amplitude decreases steadily with frequency.

The model chosen does not account for tyre flexibility, which has been shown to reduce lateral force under transient conditions. At 50 mile/h the upper frequency of 5 Hz gives a wavelength for tyre movement over the road of 4.47 m. This is aproaching the condition in which lateral force will be significantly reduced. It is therefore to be expected that some further reduction in the amplitudes of yaw velocity and side-slip will be present.

3.19 COMMENT

This chapter describes the basic vehicle and its responses to steering. The concepts developed here may be applied to any vehicle and at any time during a manoeuvre. Steering gear stiffness, lateral tyre flexibility have been examined as additions to the basic two-degree-of-freedom model. Although developed separately these modifications can be combined and when modelling the steering system it is usual to allow for tyre lateral and fore/aft deflections since these change the moments applied to the steering.

The equations of motion can be used to analyse the vheicle at any level of lateral acceleration, always provided that the appropriate slopes of the F&M curves are selected.

REFERENCES

(1) L.SEGEL (1956) *Theoretical prediction and experimental substantation of the responses of the automobile to steering control*, Institution of Mechanical Engineers, London.

(2) ISO DIS 4138.2 *Steady state circular test* International Organization for Standardisation.

(3) *Road vehicle transient response test procedure (step/ramp input)* (1979), draft proposal for an International Standard, ISO/TC22/SC9/N185.

(4) *Road vehicle transient response test procedure (sinusoidal input)* (1979), draft proposal for an International Standard, ISO/TC22/SC9/N191.

(5) *Road vehicle transient response test procedure (random input)* (1979), draft proposal for an International Standard, ISO/TC22/SC9/N194.

CHAPTER FOUR

Suspension Characteristics

4.1 INTRODUCTION

The suspension of a road vehicle is usually designed with two objectives: to isolate the vehicle body from road irregularities, and to maintain the road wheels in contact with the roadway in such a manner that the road irregularities and the body movements in bounce, pitch, and roll do not affect directional control and stability except as required by the vehicle designer.

Isolation is achieved by the use of springs and dampers to suspend the vehicle body as a seismic mass against the action of long wave length disturbances and by rubber mountings at the connections of the individual suspension components to the body for shorter wave length, or higher frequency, inputs.

Wheel control requires a mechanical guidance system which is usually a three dimensional linkage. This linkage has to maintain the position of the wheel relative to the road in all circumstances.

This chapter starts with a statement of the possible relations between the spatial movements of the vehicle body, the lateral and longitudinal tyre forces and the response of the wheel on the road. A further section gives descriptions of some actual suspensions and notes the purpose of the various parts. It is expected that this section will enable the reader to appreciate some of the problems of suspension design.Suspension analysis is introduced by considering the mechanism as a two dimensional linkage, a reasonable assumption in most cases since the majority of movement occurs in the yz plane through the centre of the road wheels. A three dimensional method is illustrated later. The 'roll centre','suspension derivative', and 'finite displacement' methods of analysis are discussed; the first method is most suitable for graphic analysis while the other techniques are best solved by hand calculator or computer.

4.2 SUSPENSION PARAMETERS

The quasi-static performance of a suspension can be defined by five variables.

(1) Normal force between tyre and road.
(2) Fore/aft movement of tyre on surface.
(3) Lateral movement of tyre.
(4) Steer rotation.
(5) Camber rotation.

4.3 SUSPENSION CHARACTERISTICS

The characteristics of a suspension are the changes in the suspension parameters caused by movements of the vehicle and forces in the plane of the road. The characteristics can be divided into two categories: the *kinematic characteristics* caused by the bounce, roll, and pitch of the body; and the *compliance characteristics* which are elastic deformations of the suspension mechanism due to fore/aft force, lateral force, and a moment in the plane of the road.

Kinematic characteristics

Inputs	bounce, roll, pitch.
Linear responses	fore/aft and lateral tyre movement.
Rotary responses	steer and camber angles.
Force response	change in normal force.

In the kinematic mode the suspension is considered simply as a mechanism with a reaction onto the frictionless plane of the road. The normal force is due to the spring contained in the suspension. The list of possible body movements and the resulting wheel displacements and normal force is given above.

Forces and moments generated by the tyre on the road distort the suspension from the position given by the kinematic mode; this elasticity of the mechanism is *compliance*. A majority of the compliance appears to come from the rubber bushings which are used to separate the suspension from the body. Compliance may be used to advantage to develop tyre attitudes which lead to some desired change in tyre forces at the upper limits of responses.

Compliance characteristics

Inputs	road-way forces and moments X, Y, N.
Linear responses	fore/aft and lateral tyre movement.
Rotary responses	steer and camber angles.
Force response	change in normal force.

These characteristics do not account for the action of the damper which develops force as a reaction to relative velocity along the line of stroke of the unit. Suspension tests measure the displacement of the ends of the damper during the kinematic mode and when applied to simulations, combine those data with the independently measured damper force/velocity characteristics to generate changes in normal force due to velocity.

4.4 SOME SUSPENSION TYPES

The realisation of a suspension linkage demands an understanding of the forces involved and the required motion of the tyre contact patch on the road. This section contains sketches of some typical designs, with comment.

At the contact between each tyre and the road it is possible that forces are generated in the three directions, *x, y, z*. A moment around the *z* axis is also to be expected; these forces and the moment correspond to the tyre forces discussed in Chapter One.

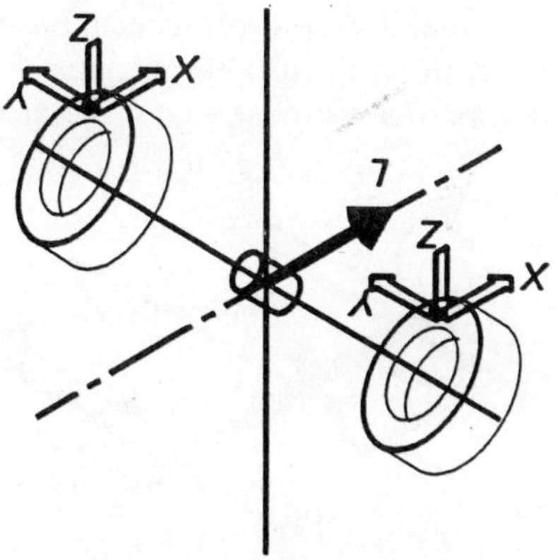

Fig.4.1 Forces and moments acting on a suspension

The forces and moments are designated $X_{l,r}$, $Y_{l,r}$, in Fig.4.1 and are shown acting in the positive sense. The drive line applies a torsion, *L*, in the line of the drive, for most purposes this may be taken as parallel to the *x* axis of the vehicle. Unless other arrangements are made reactions to the driveline torque will appear as changes in the normal forces at the wheel/road contact areas. One of the major advantages of independent suspensions for driven axles is that the driveline torque is contained by mounting the gear casing for the right angle drive and differential onto the vehicle frame thus providing a closed path for the torque.

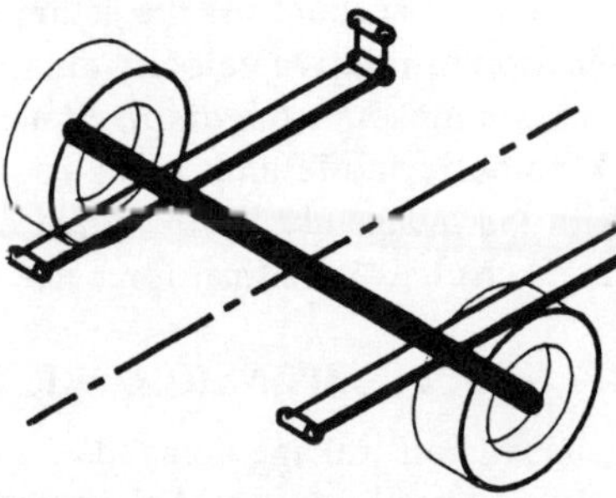

Fig 4.2(a) A simple beam axle and leaf spring layout

The basic beam axle layout is shown in Fig.4.2(a). Examination of this suspension in conjunction with the loading diagram shows that the leaf springs are required to guide the road wheels in the vertical direction, carry lateral loads, and resist the moments applied by longitudinal forces in addition to their proper function as a spring medium. These demands may distort the suspension in an unsuitable manner and the arrangement requires some modification if a well tempered suspension is to be designed.

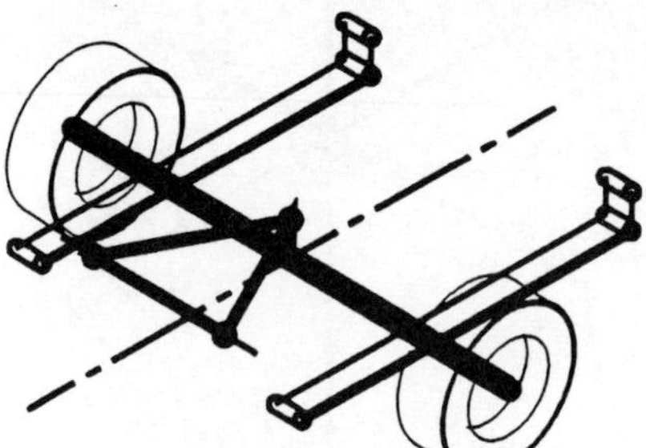

Fig.4.2(b) The triangular member relieves the leaf springs of lateral forces and prevents axle rotation in braking

The performance of the leaf spring can be improved by guiding it with links located to releive the springs of all functions other than springlike support. The purpose of the triangulated link shown in Fig.4.2(b) is to provide lateral location and resistance to rotation of the axle under braking. Two further examples of this guidance are shown in the sketches (c) and (d). In (c) the fore/aft links are torsionally flexible, while the lateral link, or 'panhard rod' provides lateral location.

The beam axle is frequently used at the the rear of front drive cars where its ability to maintain the wheels perpendicular to the road is important. A typical layout is shown in Fig.4.2(c) where the longitudinal arms connecting the axle to the frame are torsionally and laterally flexible so that a mechanism is obtained in both bounce and roll. Wind up under braking is eliminated by attaching the longitudinal arms to the axle so that relative rotation is not possible. Lateral deflection is resisted by the tie rod which has one end attached to the axle and the other to the frame with revolute, or ball, joints. The springing is by coil springs.

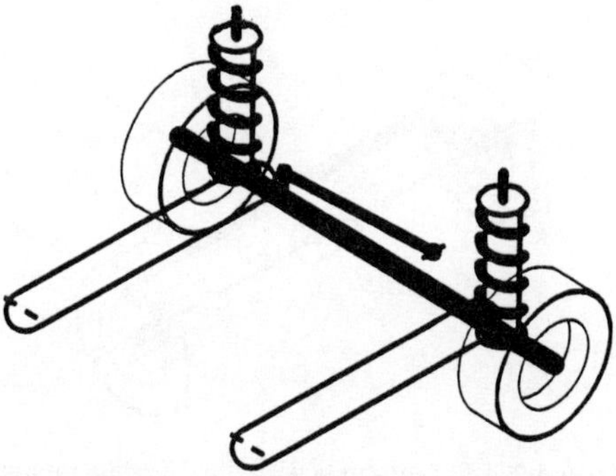

Fig.4.2(c) A method of locating the beam axle. The trailing arms are torsionally flexible

The Watt's linkage is a mechanism designed to give straight line motion. Two linkages are incorporated with a lateral rod in the beam axle control mechanism shown in Fig.4.2(d).

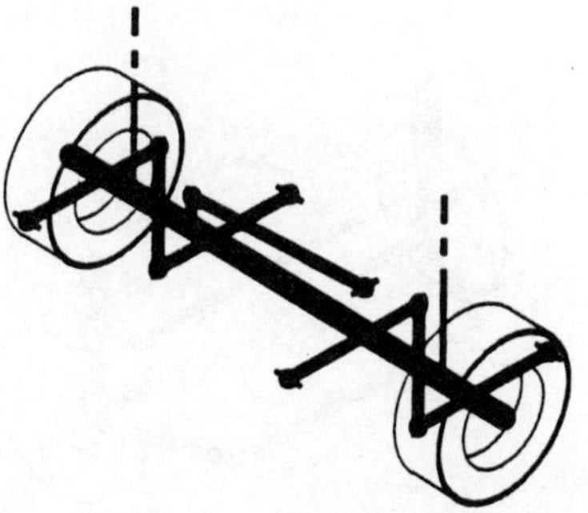

Fig.4.2(d) Another location method using Watt's linkage

The 'De Dion' design shown in Fig.4.2(e) is a modification of the beam axle in which the transmission torque is isolated from the road wheels. It is sometimes considered that the ability of the beam axle to maintain the wheels in an upright position at all times is worth the complication of a design which mounts the rear axle transfer box on the frame and employs universal joints and splined shafts to drive the wheels, thus ensuring that the normal forces between the tyre and the road are not affected by the drive line torque. Neither the locating links or the springs and dampers are shown. A system of control arms similar to that shown in Fig.4.2(c) is one possibility.

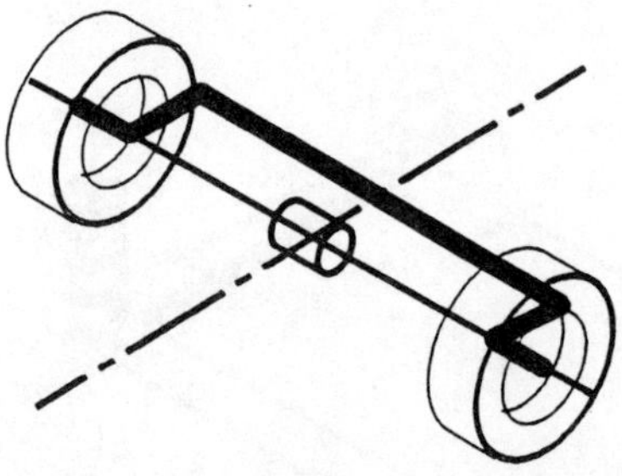

Fig.4.2(e) The De Dion axle is a beam axle with the driveline mounted on the chassis

Independent suspensions in which one wheel can move without affecting the opposite wheel take many forms. The strut and 'short/long arm' suspensions are inversions of the same mechanism. The 'short/long arm' suspension shows that five links are required to control wheel motion, although in many designs two links will be combined as a single pressing.

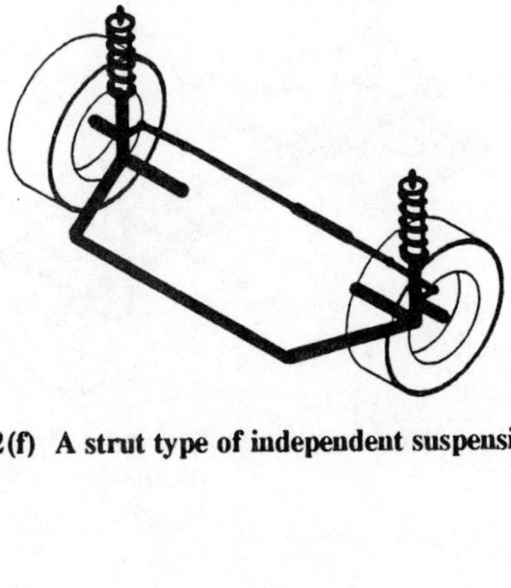

Fig.4.2(f) A strut type of independent suspension

A single pivotted member to which the wheel hub is rigidly attached can be aligned with the pivot axis parallel to the longitudinal axis of the vehicle, Fig.4.3(a), to form a 'swing axle'. The swing axle suspension is simple but the design is notorious for its habit of raising the vehicle under the action of unequal lateral tyre forces. In some designs the revolute joints between the links and the frame are mounted on the far side of the frame from the wheel with the intent of providing a longer arm, thus reducing the tendency to 'jacking'.

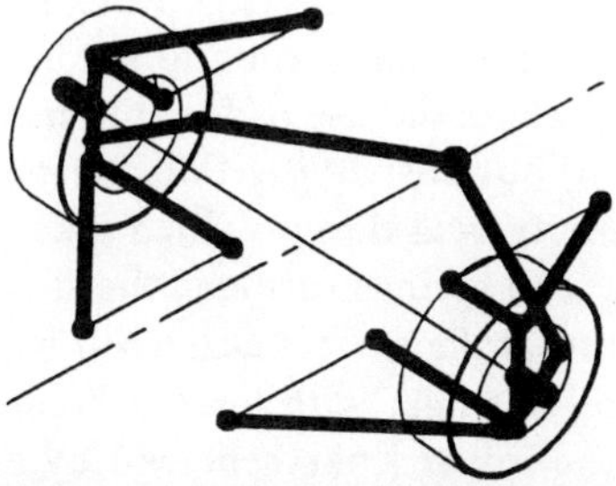

Fig.4.2(g) The 'short/long arm' linkage uses five links

The trailing arm design shown in Fig.4.3(b) causes the wheel to move in an arc around the pivot parallel to the y axis of the vehicle. Thus roll of the vehicle body results in the wheel adopting a similar angle of camber. This layout is used as the rear suspension in a number of front drive sub-compact cars, the design can be accomodated in a small space and provides a low platform height at the rear.

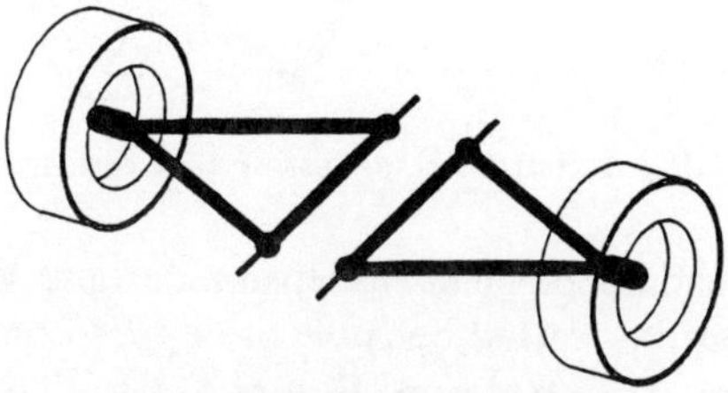

Fig.4.3(a) The swing axle

A compromise between the swing axle and the trailing arm design is provided by the semi-trailing arm shown in Fig.4.3(c). The pivot axis may be inclined to vary the handling responses by changing the length of the 'equivalent swing axle'. This design appears on a number of rear drive cars.

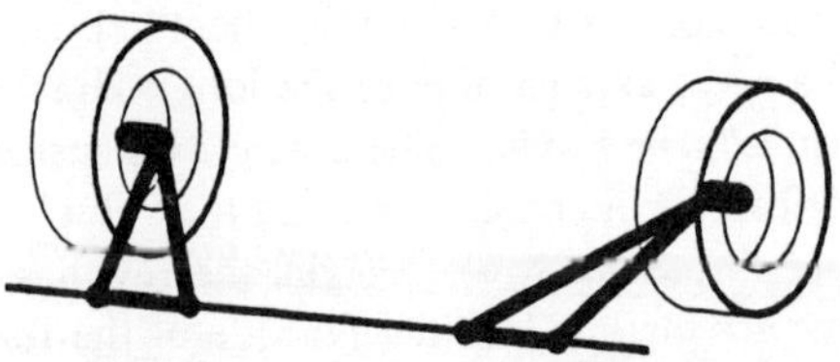

Fig.4.3(b) Trailing arm suspension

The tripod layout of the strut suspension provides a simple and effective means of force tranfer from the suspension to the body. The well separated mounting points also permits the use of flexible mountings which reduce noise transmission without destroying the original kinematic concept. Figure 4.2(f) shows the typical design which may be used at the rear or front of a vehicle. The coil springs surround the struts. Note that the crank link joining the two struts at the front is also a roll restraint. This design has been executed with coil springs, with torsion bars attached to the lower links and even with the lower links replaced by a continuous flexible member which is clamped at the inner mounting points and provides a means of modifying roll and bounce stiffnesses independently.

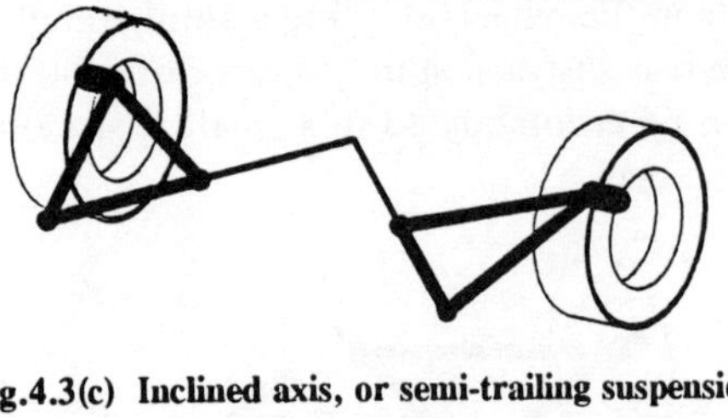

Fig.4.3(c) Inclined axis, or semi-trailing suspension

One of the earliest independent suspension designs was the 'short/long arm' (SLA), system, so called because the upper control arm is much shorter than the lower control arm. Figure 4.2.g). The term 'five link' suspension is also used. This design is not much used on production vehicles because the links intrude into the baggage compartment; however, some specialist cars employ a rear suspension of this type. A variation of the suspension used at the rear with upper and lower arms of approximately equal length employs the drive shaft between the wheel and the differential as the upper arm of the linkage, with the lower arm attached to the frame.

This design does not intrude upon baggage space; the bearings of the differential carry the lateral forces from the tyres.

The typical front suspension is usually either a strut or SLA, as both designs can be arranged not to intrude into the engine compartment.

4.5 SUSPENSION ROLL CENTRE

For a two dimensional representation of a suspension there exists an instantaneous centre for which it is possible to rotate the body cross section while the points of tyre/road contact do not slide. This centre is the *roll centre*. Although this artificial point has no real existence, it is helpful in elementary suspension design.

Consider a simple suspension linkage which can be adequately represented by a mechanism in a *yz* plane through the centres of the road wheels which comprise the axle set. The roll centre is located by placing revolute joints at each of the wheel/road contact points and employing the concept of relative instantaneous centres.

The suspension consists of two links connecting the wheel hub assembly to the body; the wheel is replaced by a triangulated linkage. The body, each link and earth are numbered as shown in Fig.4.4. The relative instantaneous centres between the parts are the revolute joints.

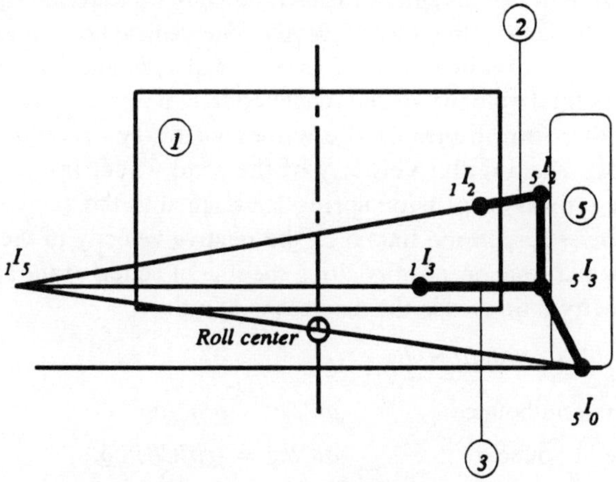

Fig.4.4 Location of the instantaneous roll centre

Body - upper link	${}_{1}I_{2}$
Body - lower link	${}_{1}I_{3}$
Wheel - upper link	${}_{5}I_{2}$
Wheel - lower link	${}_{5}I_{3}$
Wheel - earth	${}_{5}I_{0}$

From the three centres in line theorem the instaneous centre of the wheel to body, ${}_{1}I_{5}$, is located at the intersection of lines ${}_{1}I_{2}$ - ${}_{2}I_{5}$ and ${}_{1}I_{3}$ - ${}_{3}I_{5}$ extended as necessary. ${}_{1}I_{0}$ the centre of the body to earth, lies along the line through ${}_{1}I_{5}$ and the wheel/earth instantaneous centre ${}_{0}I_{5}$. By symmetry this final centre, the roll centre, is on the centreline of the vehicle. It is clear that once the vehicle departs from the erect position that symmetry no longer exists and the roll centre moves from the centreline. The roll centre will be employed in a graphical analysis which enables roll stiffness to be estimated.

4.6 ANALYSIS OF A SIMPLE SUSPENSION USING THE 'ROLL CENTRE'

The velocity diagrams for the bounce and roll modes are constructed. In the bounce diagram it is convenient to start the velocity diagram by imparting unit angular velocity to the upper link *AD*. The vehicle body does not move, hence the velocity of joints *A* and *B* is zero, and *a, b,* and *o* are coincident.

$\underline{oo}'$ is the total velocity of the wheel contact point, *O'*. The horizontal and vertical components of the wheel velocity are $\underline{o'o}''$ and $\underline{oo}''$, respectively. The angular velocity of the road wheel in the *yz* plane is $\underline{cb}/CB$. The typical suspension spring is attached to the vehicle body at *F* and to the lower suspension link at *E*, the relative velocity of these points is $\underline{ef}$, and $\underline{ef}'$ is the component of $\underline{ef}$ along the line of action of the spring. Note that the velocity ratio is also the displacement ratio.

$$dl/dt = (dl/dz)/(dz/dt)$$

Tyre scrub rate in bounce: $dy'/dz = \underline{o'o}''/\underline{oo}''$

Camber rate in bounce: $d\phi'/dz = (\underline{cb}/CB)/\underline{oo}''$

Spring motion in bounce: $dl/dz = \underline{ef}'/\underline{oo}''$

From *o*	<u>*ad*</u> is perpendicular to *AD*. Known length
From *o*	<u>*bc*</u> is perpendicular to *BC*
From *b*	<u>*dc*</u> is perpendicular tp *DC*; intersects *bc* at *c*
From *b*	<u>*bo'*</u> is perpendicualr to *BO'*
lh 2	
From *c*	<u>*co'*</u> is perpendicular to *CO'*; locates *o'*

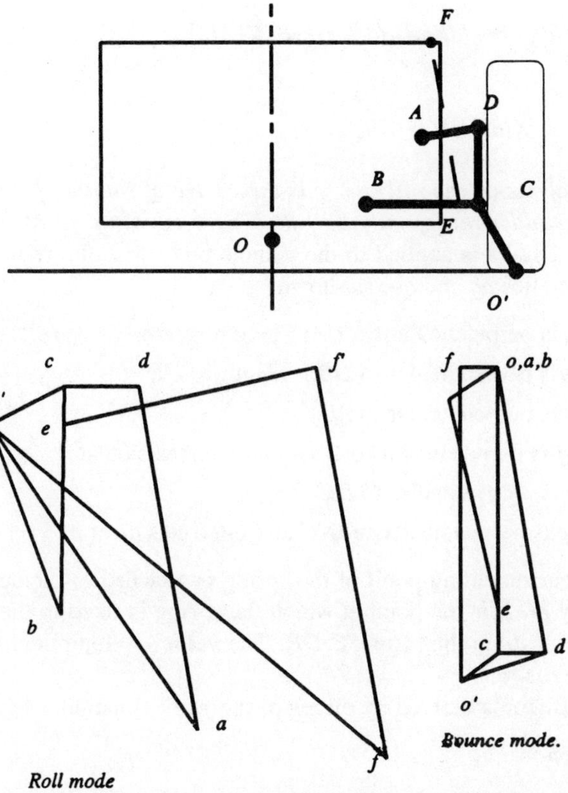

Fig.4.5 Velocity diagrams for the SLA suspension. These diagrams use the 'roll centre' so that the tyre contact patch is a revolute joint *O'*

From the change in length of the spring and its stiffness the effective spring rate of the suspension at the wheel/road interface, *the bounce stiffness*, is determined by means of the work equation.

External work = Work done by suspension spring

$$Zdz = Fdl$$

$$Z = Fdl/dz$$

Then

$$dZ/dz = (dF/dz)dl/dz + F(d^2l/dz^2)$$

Where

$$dF/dz = (dF/dl)/(dl/dz) = K\ dl/dz$$

Hence

$$K_z = K(dl/dz)^2 + F(d^2l/dz^2) \qquad (4.1)$$

In the roll mode *the roll centre is the centre of rotation for the body, so the wheel/road contact may be taken as a revolute joint.* An angular velocity of 1 rad/s is applied to the vehicle body with the result that *A* and *D* have velocities *oa* and *ob*, as shown.

$\underline{oa}$ is perpendicular to *OA*; length = *OA*

$\underline{ob}$ is perpendicular to *OB*; length = *OB*

$\underline{ad}$ is perpendicular to *AD*

$\underline{o'd}$ is perpendicular to *O'D* and intersects *ad* at *d*

$\underline{dc}$ is perpendicular to *DC*

$\underline{o'c}$ is perpendicular to *O'C* and intersects *dc* at *c*

The upper mounting point of the spring is attached to the body at *F* and has velocity $\underline{of}$. *E* is the point at which the spring is fixed to the lower link, thus $\underline{e}$ divides $\underline{dc}$ in the ratio *CE:DE*. The velocity along the line of action of the spring is $\underline{fe}$.

Roll stiffness is derived by means of the work equation.

$$Ld\phi = Fdl$$

$$L = Fdl/d\phi$$

$$dL/d\phi = (dF/d\phi)dl/d\phi + F(d^2l/d\phi^2)$$

hence

$$K_\phi = K(dl/d\phi)^2 + F(d^2l/d\phi^2) \qquad (4.2)$$

The roll centre is specified only for the zero roll angle and the graphical solution applies only to that position.

4.7 SUSPENSION DERIVATIVE ANALYSIS

By using a vectorial approach an analysis is developed in terms of the body bounce and roll motions as inputs to the suspension mechanism for which the outputs are the wheel lateral motion and camber velocity.

This analysis has the advantages that it is not dependent upon the location of a roll centre and that it can be applied at any vehicle attitude.In Fig.4.6 the origin O is fixed in the vehicle body (inboard point) while the point O' is attached to the wheel (outboard point) at the point of contact with the road. The body is permitted bounce and roll velocities while the wheel has lateral and camber movements. The body origin may be at any position in the plane; it is shown here in the central position at ground level.

This position is selected because it corresponds to the position used in the suspension parameter measurement machines designed by the author and now in operation in various test laboratories.

The analysis will consider the velocity of both the 'inboard' and 'outboard' points in terms of the body and wheel velocities respectively and then relate these velocities by their connection through the rigid link joining the two points.

The coordinates of the inboard point are (y,z); the velocity is developed from Fig.4.6.

$$V_a = (-pz)j + (W+py)k \qquad (4.3)$$

The coordinates of the outboard point, attached to the wheel, are (y',z') relative to the wheel and (y,z) relative to the body origin.

$$y' = y - y_{O'}; \quad z' = z - z_{O'}$$

The velocity of the outboard point is composed of the velocity of the wheel, V', and a component p' due to the rate of camber rotation of the wheel.

$$V_d = (V'-p'z')j + p'k \qquad (4.4)$$

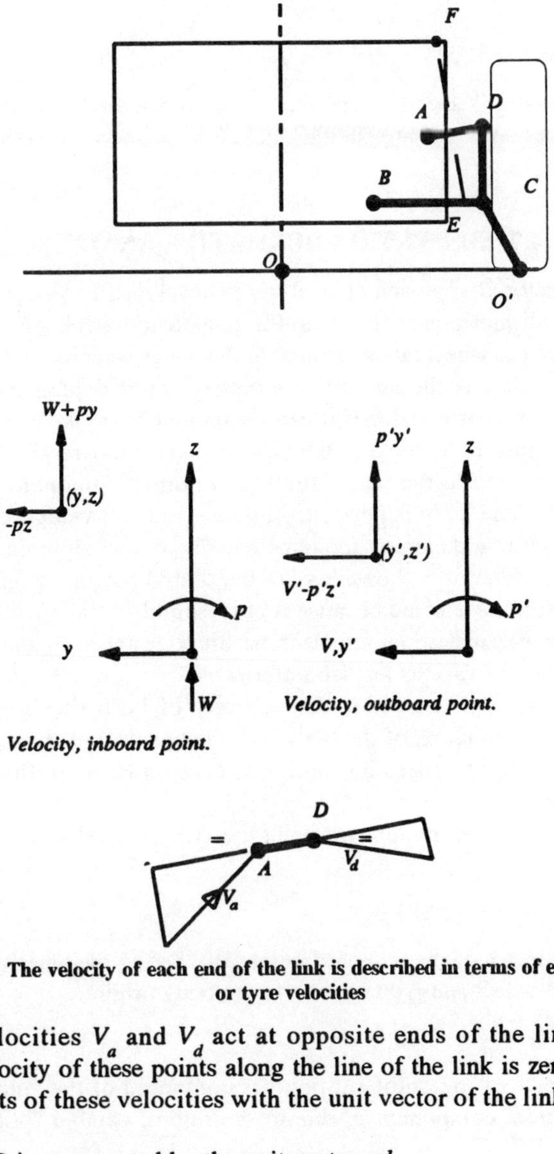

Fig.4.6 The velocity of each end of the link is described in terms of either body or tyre velocities

The velocities V_a and V_d act at opposite ends of the link AD. The relative velocity of these points along the line of the link is zero, hence the dot products of these velocities with the unit vector of the link must equal each other.

Link *AD* is represented by the unit vector $\underline{ad}$.

$$\underline{ad} = mj + nk \tag{4.5}$$

where

$$L^2 = (y_a - y_d)^2 + (z_a - z_d)^2$$

$$m = (y_a - y_d)/L$$

$$n = (z_a - z_d)/L$$

The velocities of the body and at the wheel/road interface are related by the following equation.

$$(-pz_d)m + (W + py_d)n = (V' - p'z'_d)m + p'y'_d n \qquad (4.6)$$

Each link connecting the vehicle body and the wheel yields a similar equation with the result that is shown in equation (4.8) when applied to the suspension shown in the upper part of Fig.4.7.

$$\begin{bmatrix} m_1 & -(z_d - z_{0'})m_1 + (y_d - y_{0'})n_1 \\ m_2 & -(z_c - z_{0'})m_2 + (y_c - y_{0'})n_2 \end{bmatrix} \begin{bmatrix} V' \\ p' \end{bmatrix} = \begin{bmatrix} n_1 & -z_a m_1 + y_a n_1 \\ n_2 & -z_b m_2 + y_b n_2 \end{bmatrix} \begin{bmatrix} W \\ p \end{bmatrix} \qquad (4.7)$$

The most convenient method of solution is by Cramer's Rule. First check that the Jacobean is not zero.

$$\begin{vmatrix} m_1 & -(z_d - z_{0'})m_1 + (y_d - y_{0'})n_1 \\ m_2 & -(z_c - z_{0'})m_2 + (y_c - y_{0'})n_2 \end{vmatrix} \neq 0$$

The ratio between lateral velocity of the contact patch and bounce velocity is then obtained.

$$V'/W = \frac{\begin{vmatrix} n_1 & -(z_d - z_{0'})m_1 + (y_d - y_{0'})n_1 \\ n_2 & -(z_c - z_{0'})m_2 + (y_c - y_{0'})n_2 \end{vmatrix}}{\begin{vmatrix} m_1 & -(z_d - z_{0'})m_1 + (y_d - y_{0'})n_1 \\ m_2 & -(z_c - z_{0'})m_2 + (y_c - y_{0'})n_2 \end{vmatrix}}$$

The tyre scrub due to roll is

$$V'/p = \frac{\begin{vmatrix} -z_a m_1 + y_a n_1 & -(z_d - z_{0'})m_1 + (y_d - y_{0'})n_1 \\ -z_b m_2 + y_b n_2 & -(z_c - z_{0'})m_2 + (y_c - y_{0'})n_2 \end{vmatrix}}{\begin{vmatrix} m_1 & -(z_d - z_{0'})m_1 + (y_d - y_{0'})n_1 \\ m_2 & -(z_c - z_{0'})m_2 + (y_c - y_{0'})n_2 \end{vmatrix}}$$

Camber due to bounce and roll are given below

$$p'/W = \frac{\begin{vmatrix} m_1 & n_1 \\ m_2 & n_2 \end{vmatrix}}{\begin{vmatrix} m_1 & -(z_d - z_{0'})m_1 + (y_d - y_{0'})n_1 \\ m_2 & -(z_c - z_{0'})m_2 + (y_c - y_{0'})n_2 \end{vmatrix}}$$

$$p'/p = \frac{\begin{vmatrix} m_1 & -z_a m_1 + y_a n_1 \\ m_2 & -z_b m_2 + y_b n_2 \end{vmatrix}}{\begin{vmatrix} m_1 & -(z_d - z_{0'})m_1 + (y_d - y_{0'})n_1 \\ m_2 & -(z_c - z_{0'})m_2 + (y_c - y_{0'})n_2 \end{vmatrix}}$$

The ratios V'/W, p'/W, V'/p and p'/p which are given above are also the rates of change of position for the suspension since, as noted previously:

$$dV'/dW = (dy'/dt)/(dz/dt) = dy'/dz, \quad \text{etc.}$$

These ratios are the *suspension derivatives*, and describe the possible movements of the wheel on the road as the vehicle body rolls and bounces. Note that the derivatives do not depend upon the location of an arbitrary point and the method may be applied for any roll angle or vertical displacement, or combination of these inputs.

The suspension spring may be a torsion bar or a coil spring for example. In either case it is usually attached to the lower link and the body. Consider the case of an arbitrary point, E, not colinear with B and C. The velocity of this point is, with v_e and w_e as the components in the y and z directions

$$v_e = v_b + (v_c - v_b)(z_e - z_b)/(z_c - z_b)$$
$$w_e = w_b + (w_c - w_b)(y_e - y_b)/(y_c - y_b) \qquad (4.9)$$

The velocity of the spring mounting point, *F*, is that of the typical point on the body.

$$V_f = -pz_f; \quad W_f = W + py_f$$

The direction of the spring is given by the direction cosines m_3 and n_3 which are derived from the positions of points *E* and *F*.

$$\underline{ef} = m_3 j + n_3 k; \quad m_3 = (y_f - y_e)/L \quad n_3 = (z_f - z_e)/L$$

The relative velocity of the two ends of the spring is obtained as the difference in the velocities of *E* and *F* along the line of the spring. That is of the difference in the dot products of these velocities with the unit vector of the spring direction.

$$dl/dt = (v_e + pz_f)m_3 + (w_e - (W+py_f))n_3$$

Note that when the link is horizontal or vertical one of the above terms will give a 'divide by zero' error when programmed. Thus it is prudent to take steps to avoid this in the coding.

Effective spring stiffness in roll and bounce referred to the road contact point is obtained from equations (4.1) and (4.2), respectively. This analysis is directly applicable to the determination of the additional bounce and roll stiffnesses provided by auxiliary springs such as anti-sway bars and bounce bars which are used to provide additional roll stiffness without affecting bounce stiffness or the converse.

Analysis of suspensions with sliding members, e.g., the strut suspension is facilitated if the sliding member is replaced by an infinitely long link. This link is perpendicular to the direction of sliding and hence if the direction cosines of the sliding member are m_1 and n_1 then the direction cosines of the link are n_1 and $-m_1$. Note the reversal of the cosines and the sign change. Otherwise the analysis is as before, the spring now resides between the lower ball-joint and the body mount and the computation of the closing velocity is thus simplified.

4.8 THE BEAM AXLE

The characteristics of the beam axle may be determined in those cases where a lateral rod couples the axle to the vehicle body, in other cases it may be necessary to assume such a link. Figure 4.7. shows an idealised beam axle with a lateral locating link. The analysis follows the suspension derivative method.

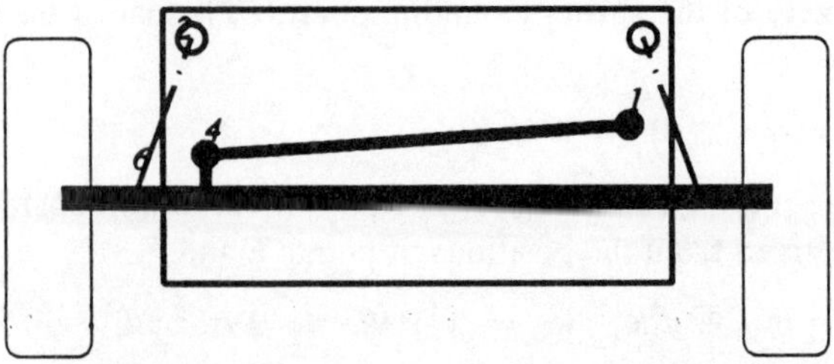

Fig.4.7 The characteristics of the beam axle suspension

$$V_1 = (-pz_1)j + (W+py_1)k$$

The axle cannot roll unless the tyres deform, and thus the only motion to be considered is the lateral velocity

$$V_4 = (V')j$$

The link is represented by the unit vector. Equate the velocity components of points 1 and 4 along the link.

The link vector is

$$\underline{1\text{-}4} = m.j + n.k$$

Hence

$$(-pz_1)m + (W+py_1)n = V'm$$

Consider the bounce and roll separately to obtain the characteristics of the suspension.

Roll: $dy'/d\phi = -z_1 + y_1.n/m$ (4.10)

When the link is horizontal, $n = 0$, $m = 1$, and the characteristic becomes for roll: $dy'/d\phi = -z_1$

in bounce: $dy'/dz = n/m$ (4.11)

For the general case of an inclined spring the velocities of the mounting points of the spring are resolved in the direction of the line joining the spring centres.

$$V_2 = (-pz_2)m_2 + (W+py_2)n_2$$

The subscript 2 for the direction cosines refers to the line of action of the spring.

$$V_6 = V'j$$

$V'$is known from the previous calculation, hence the relative velocity of the spring mounts is obtained.

$${}_2V_6 = (-pz_2)m_2 + (W+py_2)n - V'm_2 \qquad (4.12)$$

This calculation also provides the relative velocity across the dampers in the general case.

4.9 SOME CHARACTERISTICS OF LEAF SPRINGS

Leaf springs are frequently used in trucks and the roll stiffness of the suspension is affected by the design and position of the axle mounts on the springs.

In general the application of a load to a beam results in both a deflection and a rotation at the point of loading. Conversely when a moment is applied then a rotation and deflection occur at the loaded section. These effects are linked by the 'reciprocal theorem' of mechanics and the *change in slope produced by an applied unit force is the same as the deflection at the loading cross section caused by an unit moment.*

Thus in the roll mode when two leaf springs are linked by the beam axle one spring deflects upward while the other deflects down. Each deflection is accompanied by a tendency to change in slope which is nullified by the presence of the axle. Figure 4.8 shows the deflected modes of a single leaf spring.

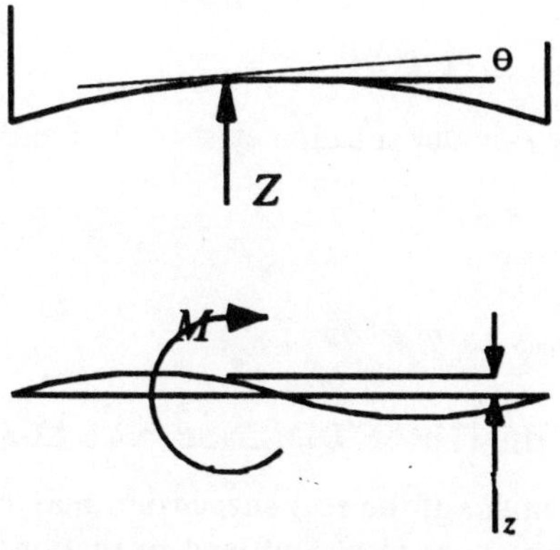

Fig.4.8 The deflection and rotation of the leaf spring

The relationships shown in Fig. 4.8 may obtained either by test or computation.

The effect of restraining these rotations is determined using the work equation. K_Θ is the resistance of the leaf spring to a unit moment.

$$2T = K_z z_1^2 + K_z z_2^2 + K_\Theta(\Theta_1 - \Theta_2)^2$$

$$\Theta_1 = d\Theta/dz.z_1; \quad \Theta_2 = d\Theta/dz.z_2$$

$$z_1 = -y_s\phi; \quad z_2 = y_s\phi$$

$$T = (K_z + K_\Theta d^2\Theta/dz^2)(y_s\phi)^2$$

but $dT/d\phi = K_\phi \phi$ and the roll stiffness is obtained.

$$K_\phi = 2(K_z + K_\Theta d^2\Theta/dz^2)y_s^2 \tag{4.13}$$

4.10 THE EFFECTS OF TYRE STIFFNESS

From the previous work the effective bounce and roll stiffnesses of a suspension may be calculated. It is frequently useful to consider the suspension and tyre stiffnesses together and the necessary equations are now developed.

Bounce: $z = z_s + z_t$

The force acting on both springs is the same

$$z = Z/K_z + Z/K'_z$$

Hence

$$K = dZ/dz = K_z K'_z/(K_z + K'_z) \tag{4.14}$$

For the roll mode a similar situation exists.

$$\phi = L/K_\phi + L/K'_\phi$$

Hence

$$K = dL/d\phi = K_\phi K'_\phi/(K_\phi + K'_\phi) \tag{4.15}$$

4.11 THE THREE DIMENSIONAL SUSPENSION

The spatial mechanism of the real suspension may be analysed using the suspension derivative approach outlined in section 4.7. The suspension parameters for the suspension are:

(a) Longitudinal movement of the tyre on the road.
(b) Lateral movement of the tyre on the road.
(c) Steer rotation.
(d) Camber rotation.
(e) Normal force of contact.

The body movements which cause relative motion of the tyre on the road are:

(a) Bounce.
(b) Roll.
(c) Pitch.

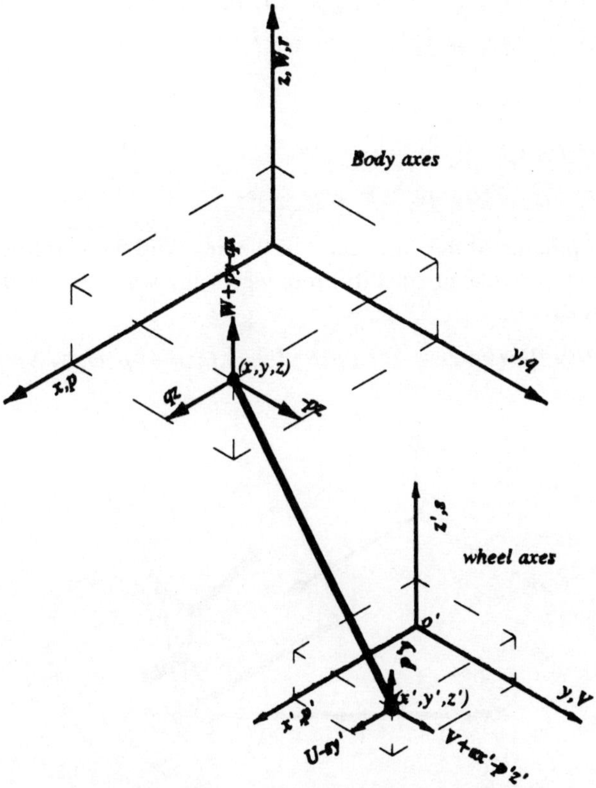

Fig.4.9 The velocities of a link attached to the vehicle frame (1) and the stub axle (2) for a three dimensional suspension model

As in the two dimensional analysis the velocity of a point attached to the vehicle body and a second on the stub axle are established.

The velocity of an inboard point is as shown in Fig.4.9

$$V_1 = (qz)i + (-pz)j + (W + py - qx)k$$

The typical point attached to the road wheel has the following velocity.

$$V_2 = (U - sy')i + (V + sx' - p'z')j + (p'y')k$$

Note that

$$x' = x - x_{0'}\ ;\quad y' = y - y_{0'}\ ;\quad z' = z - z_{0'}$$

The link connecting these points, $\underline{1\text{-}2}$, may be expressed as a unit vector.

$$\underline{1\text{-}2} = \underline{l}\,.\,i + m\,.\,j + n\,.\,k$$

where

$$\underline{l} = (x_2 - x_1)/L\ ;\quad m = (y_2 - y_2)/L\ ;\quad n = (z_2 - z_1)/L$$

$$L^2 = (x_2 - x_1)^2 + (y_1 - y_2)^2 + (z_2 - z_1)^2$$

The dot product of the unit vector of the link with the velocities at *1* and *2* respectively provide an equality from which the wheel and body motions may be related.

$$(U - sy')\underline{l} + (V + sx' - p'z')m + (p'y')n = (qz)\underline{l} + (-pz)m + (W + py - qx)n$$

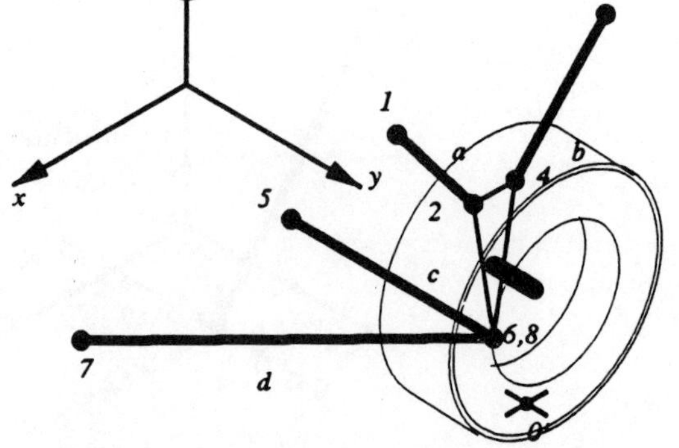

Fig.4.10 The nodes of an SLA suspension, using suspension derivative analysis as in equations (4.16)

The typical suspension can be represented by four links connecting the wheel to the body - Fig.4.10. shows a short/long arm rear suspension. The body mountings will have odd subscripts while the wheel connections will be even numbered. Note that the wheel carrier has three mounting points and the top point is No.2 with respect to the link 1-2, and No.4 for link 3-4.

The unit vector of each link is written in the following form.

Link 1,2: $\underline{1\text{-}2} = \underline{l}_a \,.\, i + m_a \,.\, j + n_a \,.\, k$
Link 3,4: $\underline{3\text{-}4} = \underline{l}_b \,.\, i + m_b \,.\, j + n_b \,.\, k$
Link 5,6: $\underline{5\text{-}6} = \underline{l}_c \,.\, i + m_c \,.\, j + n_c \,.\, k$
Link 7,8: $\underline{7\text{-}8} = \underline{l}_d \,.\, i + m_d \,.\, j + n_d \,.\, k$

The characteristic equation for the suspension is then given by equation (4.16)

$$\begin{bmatrix} \underline{l}_a & m_a & (-y_2'\cdot\underline{l}_a+x_2'\cdot m_a) & (-z_2'\cdot m_a+y_2'\cdot n_a) \\ \underline{l}_b & m_b & (-y_4'\cdot\underline{l}_b+x_4'\cdot m_b) & (-z_4'\cdot m_b+y_4'\cdot n_b) \\ \underline{l}_c & m_c & (-y_6'\cdot\underline{l}_c+x_6'\cdot m_c) & (-z_6'\cdot m_c+y_6'\cdot n_c) \\ \underline{l}_d & m_d & (-y_8'\cdot\underline{l}_d+x_8'\cdot m_d) & (-z_8'\cdot m_d+y_8'\cdot n_d) \end{bmatrix} \begin{bmatrix} U \\ V \\ s \\ p' \end{bmatrix}$$

$$= \begin{bmatrix} n_a \\ n_b \\ n_c \\ n_d \end{bmatrix} \begin{bmatrix} W \end{bmatrix} \begin{bmatrix} -z_1\cdot m_a+y_1\cdot n_a \\ -z_3\cdot m_b+y_3\cdot n_b \\ -z_5\cdot m_c+y_5\cdot n_c \\ -z_7\cdot m_d+y_7\cdot n_d \end{bmatrix} \begin{bmatrix} p \end{bmatrix} \begin{bmatrix} z_1\cdot\underline{l}_a-x_1\cdot n_a \\ z_3\cdot\underline{l}_b-x_3\cdot n_b \\ z_5\cdot\underline{l}_c-x_5\cdot n_c \\ z_7\cdot\underline{l}_d-x_7\cdot n_d \end{bmatrix} \begin{bmatrix} q \end{bmatrix} \qquad (4.16)$$

For a strut suspension the sliding and rotating joint at the top of the strut is replaced by a pair of imaginary links which are perpendicular to the strut. These links are located by determining the plane through the joint which is perpendicular to the strut and then finding the intersections of this plane with the x and y axes. $\underline{l}$, m, and n are the direction cosines of the strut.

The plane in which the sliding/rotating joint is located is given by the equation

$$\underline{l}x + my + nz = \underline{l}x_2 + my_2 + nz_2 \qquad (4.17)$$

The x axis intersects the plane when $y = 0$ and $z = 0$. Hence

$$x_1 = (\underline{l}\,x_2+my_2+nz_2)/\underline{l} \tag{4.18}$$

Then one of the required vectors is the unit vector of the line joining x_2, y_2, z_2 to $x_1,0,0$.

The y axis intersection with the plane is then obtained, and the second link is the unit vector of the line x_2,y_2,z_2 to $0,y_3,0$.

$$y_3 = (\underline{l}\,x_2+my_2+nz_2)/m \tag{4.19}$$

4.12 THE STEERING LINKAGE

The steering axis which may be formed by a pair of ball joints, as shown in Fig.4.11, or by a cylindrical joint, as is usual in heavy trucks, inclined to give trail and offset. The inclination means that a rotation around the steering axis causes steer and camber at the tyre contact patch.

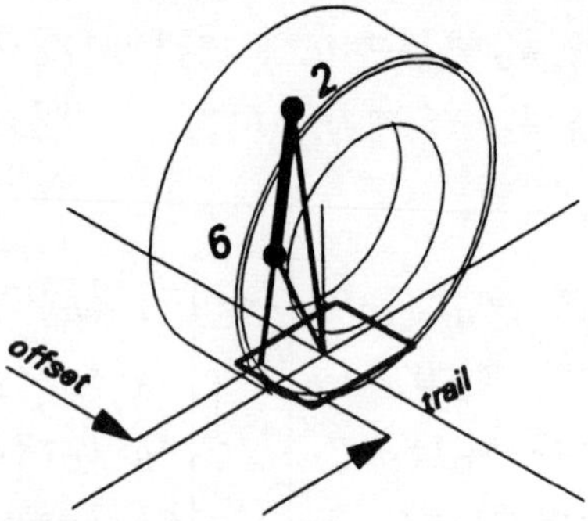

Fig.4.11 The inclination of the steering axis causes steering to have steer and camber components at the road plane

Let the steering axis vector $\underline{2\text{-}6} = \underline{l}_e i+m_e j+n_e k$, then rotation around that axis by S has the following result.

Steer angle $\quad qs = Sn_e$

Camber angle $\quad \phi' = S\underline{l}_e \qquad (4.20)$

The movement of the centre of contact of the tyre on the road is perpendicular to the vectors $\underline{0'\text{-}6}$ and $\underline{0'\text{-}2}$, thus its direction is given by the vector cross product $(\underline{0'\text{-}6})$ x $(\underline{0'\text{-}2})$.

Link 0',6: $\underline{0'\text{-}6} = \underline{l}_f \, . \, i + m_f \, . \, j + n_f \, . \, k$

Link 0',2: $\underline{0'\text{-}2} = \underline{l}_h \, . \, i + m_h \, . \, j + n_h \, . \, k$

$$A_{0'} = \begin{vmatrix} i & j & k \\ \underline{l}_f & m_f & n_f \\ \underline{l}_h & m_h & n_h \end{vmatrix}$$

The perpendicular distance of $0'$ from the steer axis may be obtained from the modulus of the vector cross product since this is twice the area of the triangle $0',2,6$. Hence if L is the distance from 2 to 6 then

$$L^2 = (x_6 - x_2)^2 + (y_6 - y_2)^2 + (z_6 - z_2)^2$$

$$h = A_{0'} / L \tag{4.21}$$

The velocity of the centre of contact due to steer rotation is the cross product of the angular velocity of steer with a vector from one of the steering joints to the ground contact, the upper steering joint is used here. This velocity will have a component along the z' axis thus showing that, in general, the act of steering changes the position of the vehicle body.

$$v = S x (\underline{0'\text{-}2}) \tag{4.22}$$

$$V_{0'} = S \begin{vmatrix} i & j & k \\ \underline{l}_e & m_e & n_e \\ \underline{l}_h & m_h & n_h \end{vmatrix}$$

$$V_{0'} = S\{(m_e n_h - m_h n_e)i + (\underline{l}_h n_e - \underline{l}_e n_h)j + (\underline{l}_e m_h - \underline{l}_h m_e)k\} \tag{4.23}$$

4.13 INITIAL ESTIMATES OF BOUNCE AND ROLL STEER

In order that the applied steer angle shall not be affected by movement of the road wheels relative to the body the ends of the steering arms must be located with care. For the two dimensional suspension it is sufficient to locate the inboard end of the drop arm at the centre of the arc swept by the outboard end when the stub axle is moved through its working path.

When the positions of the steering nodes are known then the simple plane mechanism analysis by velocity diagrams will give an estimate of bounce and roll steer. There are two extra nodes to be considered.

J is the steering node attached to the vehicle frame, and H is the wheel node. The velocity diagrams are completed in the normal way and the velocities of H and J are obtained. In the case that the joints are correctly located then vector $\underline{jh}$ will be perpendicular to the link JH. Otherwise find the velocity component across the line of action of the link JH. Let $\underline{hh'}$ be that component, then

$$ds/dz = \underline{hh'}/r \tag{4.24}$$

where r is the perpendicular from the three dimensional line of action of JH to the steering axis.

4.14 THE EFFECT OF SUSPENSION GEOMETRY ON STEERING

Steering introduces an extra degree of freedom to the suspension derivative analysis which is controlled by the steering drop arm. One end of this link is fixed in the vehicle body while the other is on the wheel. A sketch of a typical suspension is shown in Fig.4.12 An additional node, *10*, connects the wheel carrier to the steering drop arm, and node *9* is the connection of the drop arm to that part of the steering mechanism which moves with the body. The velocity of an inboard point is as shown previously in Fig.4.10

$$V_9 = (qz_9)i + (-pz_9)j + (W+py_9-qx_9)k \tag{4.25}$$

The node *10* attached to the road wheel now has velocity from the orginal computation plus a possible value from the steering axis rotation which is the cross product given below.

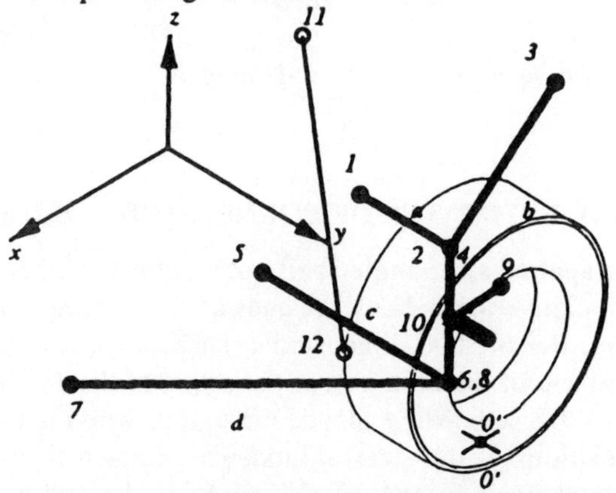

Fig.4.12 The introduction of steering in the suspension derivative analysis adds further links to the model

$$V'_{10} = \begin{vmatrix} i & j & k \\ l_e & m_e & n_e \\ l_g & m_g & n_g \end{vmatrix}$$

Hence

$$V'_{10} = S\{(m_e n_g - m_g n_e)i + (l_e n_g - l_g n_e)j + (l_e m_g - l_g m_e)k\}$$

The suspension derivative analysis will give a value of the velocity of the node attached to the road wheel for the non-steering condition. The values of U, V, s, and p' are known from a previous computation.

$$V''_{10} = (U - sy'_{10})i + (V + sx'_{10} - p'z'_{10})j + (p'y'_{10})k$$

The total possible velocity of the node is then

$$V_{10} = V'_{10} + V''_{10}$$

Let the unit vector of 9-10 be $l_p.i + m_p.j + n_p.k$, then the value of S is determined by equating the total velocities of the end points of the link along the line of the link, ie by equating dot products of the velocities and the unit vector of the link.

$$V_{10} \cdot (l_p \cdot i + m_p \cdot j + n_p \cdot k) = V_9 \cdot (l_p \cdot i + m_p \cdot j + n_p \cdot k) \quad (4.26)$$

Once the value of S is found the movements at the contact centre can be calculated from equation (4.23).

The effect of an input from the steering hand wheel may be represented by an appropriate velocity at node 9, calculation of the king pin rotational velocity from equation (4.25) and the tyre contact patch velocities from equations (4.20) and (4.22).

4.15 REACTIONS AT THE SUSPENSION MOUNTINGS AND COMPLIANCE EFFECTS

A suspension is usually isolated from the vehicle body by flexible mountings, the deflection of these mountings due to the longitudinal and lateral forces in the plane of the road changes the geometry of the suspension with the result that the characteristics of the suspension change. The change in suspension characteristics due to external forces is the compliance. A basic plane suspension is used to demonstrate a simple graphic approach. Vector methods are best used for the three dimensional suspension. The forces acting on the inboard bearings are determined by

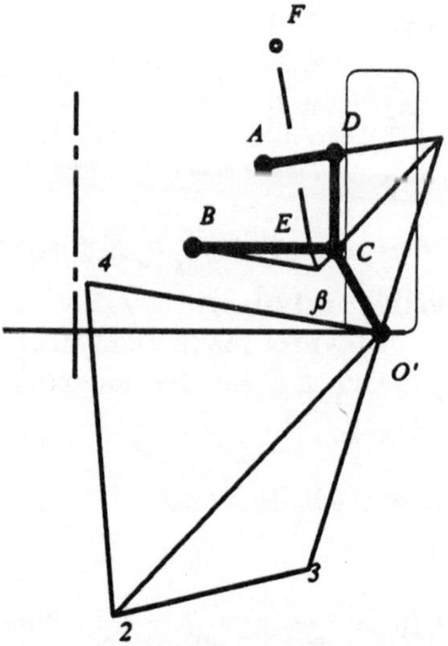

Fig.4.13 The reactions at the mounting points of a two-degree-of-freedom SLA suspension due to tyre F&M

static equilibrium. Figure 4. 13 is a suspension under the action of normal and lateral forces at the road/tyre interface, *0'*. The stub axle *0'C,D* is in equilibrium under the action of the forces at the joints.

The line of action of the force in the upper link *AD* is along *AD* and intersects the force vector at α, thus the force at the lower joint *C* also passes through α. *0'3* is the known vector, the directions of the force vectors for joints *D* and *C* are known. The lower link is now seen as under the action of the spring force along *FE* at *6* and the forces at *B* and *C*. The directions of the reaction at *C* and the spring force are known and intersect at ß, thus the direction of the reaction at *B* is obtained.

The vector diagrams *0',2,3*, and *0',2,4* give the actual forces. When the stiffness of the joints *1* and *4* is known the deflection of these joints can be determined and the new positions of the inboard nodes of the suspension derivative calculation obtained. The suspension derivative program is then used to find the changed values of *dy'/dz*, and so on.

A similar approach is followed for the three dimensional suspension, a vector analysis of forces is the preferred method, $\mathbf{F} = 0$, and $M = \mathbf{r} \times \mathbf{F} = 0$. The direction cosines of the suspension will be known from the suspension derivative program. When the suspension joints are taken as ball

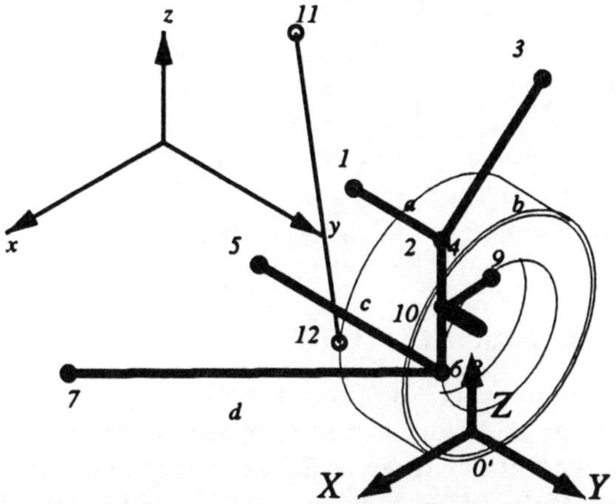

Fig.4.14 The reaction forces for the three dimensional suspension are obtained by vector methods. Note that the origin is that for the car

joints then the forces lie along the direction of the link. For example consider the link *1-2*, shown in Fig.4.14. The force at the joint is given by resolving the force in the link in directions *x, y, z* while the moment around the origin for this link is $M = \mathbf{r} \times \mathbf{F}$.

The direction cosines are l_a , m_a , and n_a

$$M = \mathbf{r} \times \mathbf{F} = \begin{vmatrix} i & j & k \\ x_1 & y_1 & z_1 \\ X & Y & Z \end{vmatrix} \tag{4.27}$$

This determinant may be expanded to give the following expression for the moment.

$$M = (Zy_1 - Yz_1)i + (Xz_1 - Zx_1)j + (Yx_1 - Xy_1)k \tag{4.28}$$

The components *X, Y, Z* of the force vector are obtained by resolving *F* along the direction of the link. Thus the previous equation becomes

$$M = F\{(m_a y_1 - m_a z_1)i + (l_a z_1 - n_a x_1)j + (m_a x_1 - l_a y_1)k\} \tag{4.29}$$

This process is repeated for each link and the spring/damper unit with the result that a 6 x 6 matrix is obtained relating the tyre contact forces to the reactions at the suspension mounting points.The matrix is shown in outline, each column should be completed with values appropriate to the nodes.

$$\begin{bmatrix} l_a \\ m_a \\ n_a \\ (m_a y_1 - m_a z_1) \\ (l_a z_1 - n_a x_1) \\ (m_a x_1 - l_a y_1) \end{bmatrix} \begin{bmatrix} F_1 \\ F_3 \\ F_5 \\ F_7 \\ F_9 \\ F_{11} \end{bmatrix} = \begin{bmatrix} X \\ Y \\ Z \\ Zy_{0'} - Yz_{0'} \\ Xz_{0'} - Zx_{0'} \\ Yx_{0'} - Xy_{0'} \end{bmatrix} \tag{4.30}$$

4.16 ROLL ATTITUDES AND LOAD TRANSFER

The steady state roll angle of a suspension is derived from consideration of potential energy and work and is expressed in terms of the suspension derivatives. Potential energy is stored in the springs by the roll of the vehicle body. Work is done by the tyres which will develop lateral forces to keep the vehicle in a turn and which move laterally on the road as a result of the roll. A single axle will be considered first and then the complete vehicle with an axle at each end.

For the case that the origin is the centre of gravity of the vehicle the roll motion of the single axle is

$$2V = K_\phi \phi^2$$

$$W = Y'y'$$

But

$$y' = dy'/d\phi . \phi$$

hence

$$W = Y'dy/d\phi . \phi$$

$$dV/d\phi = dW/d\phi$$

therefore

$$\phi = (Y'dy'/d\phi)/K_\phi \tag{4.31}$$

The forces Y' are generated at the tyres. The roll motion is due to the work done at each tyre/road contact surface, and this is resisted by the total roll stiffness, including the damping resistance in the transient mode. The torsional stiffness of the chassis is assumed to be high relative to the roll stiffness of either suspension. This is not always true, particularly with flat bed trucks.

$$\phi = \Sigma Y'dy'/d\phi/\Sigma K_\phi \tag{4.32}$$

As a vehicle moves on a curved path the lateral inertial force acting at the centre of mass gives rise to an overturning moment. This moment increases the normal force on the outer wheels and decreases the normal force on the inner wheels. Redistibution of normal force causes the lateral forces generated by the tyres moving at given attitude angles to change in a non-linear way and thus the balance of the car may change.

These effects are frequently used; for example, in rear drive cars an anti-roll or 'sway' bar may connect the two front suspensions so as to act in the roll mode and increase the transfer of normal force across the front wheels with the object of reducing the total lateral force at the front during steering and thus inducing an understeer effect.

The roll angles for the front and rear suspensions acting in isolation would be

$$\begin{aligned} \phi_f &= (Y'dy'/d\phi)_f/K_{\phi.f} \\ \phi_r &= (Y'dy'/d\phi)_r/K_{\phi.r} \end{aligned} \tag{4.33}$$

Thus the vehicle body acts as a torque tube restraining the suspensions to a common roll angle. The torque transmitted through the body may be written in terms of the front or rear suspension parameters

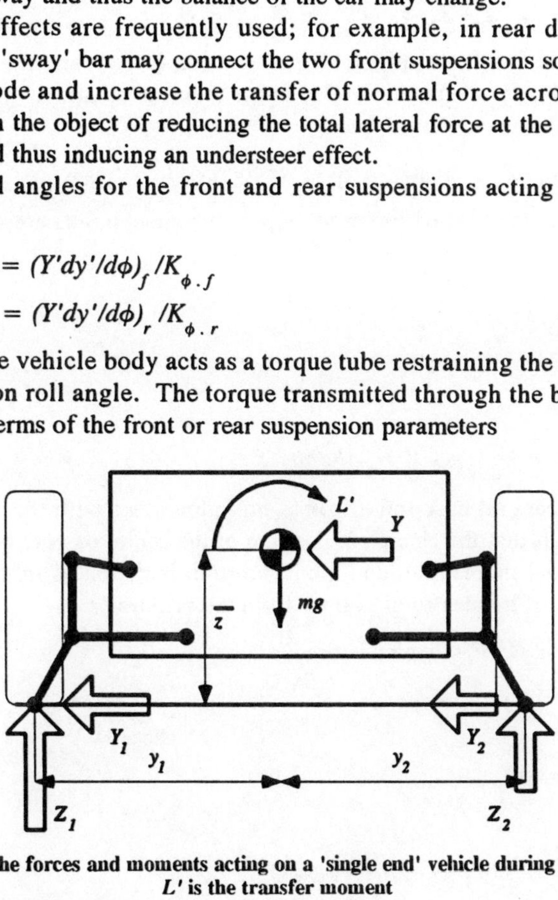

Fig.4.15 The forces and moments acting on a 'single end' vehicle during cornering. L' is the transfer moment

$$\begin{aligned} L' &= (\phi_f - \phi)K_{\phi.f} \\ L' &= (\phi - \phi_r)K_{\phi.r} \end{aligned} \tag{4.34}$$

For vertical equilibrium

$$mg + Z'_1 + Z'_2 = 0$$

Equilibrium of lateral forces

$$Y + Y'_1 + Y'_2 = 0$$

Taking moments around the origin

$$L' + Z'_2 y_2 - Z'_1 y_1 - (Y'_1 + Y'_2)\bar{z} = 0$$

Substitute for the torque, L' from equation (4.34), and the 'free' front roll angle, ϕ_f, from equation (4.33) in the moment equation.

The front suspension is typical

$$Z'_2 y_2 - Z'_1 y_1 = K_{\phi.f} \cdot \phi - Y_f (\bar{z} - dy'/d\phi_f)$$

If $y_1 = y_2$ and the normal forces at the wheel contact points are written as

$$Z_1 = mg/2 - dZ$$

and

$$Z_2 = mg/2 + dZ$$

then

$$dZ = K_{\phi.f}\phi - Y_f (\bar{z} - dy'/d\phi_f) \tag{4.35}$$

For the general case the origin is not coincident with the mass centre with the result that the change in position of the centre of gravity due to roll and the inertial moment around the new origin both appear in the equation. The position of the centre of gravity from the origin is $\bar{z}$.

$$2V = K_\phi \phi^2 - 2mg\bar{z}(1 - \cos\phi)$$

$$W = Y'y' + Y\bar{z}\phi$$

but

$$Y = -Y'$$

hence

$$\phi = Y'(dy'/d\phi - \bar{z})/(K_\phi - mg\bar{z}) \tag{4.36}$$

Note that the value of the tyre scrub with roll coefficient changes as the origin for calculation is moved vertically and equation (4.36) simply reflects this.

4.17 SUSPENSION JACKING

During bounce motion the tyres may scrub laterally on the road, this gives slip angles to the tyres with the result that forces dependent upon the bounce velocity are generated. Work is done by the movement of the tyre forces. Following the previous example it can be shown that when a vehicle is in a turn with lateral forces generated at the tyres then there may be a change in the static height of any axle.

$$V = K_z z^2/2 - mgz$$

$$W = Ydy'/dz \,.\, z$$

$$z = (Y'dy'/dz)/K_z \tag{4.37}$$

Clearly the values of dy'/dz for the two wheels on an axle are of opposite sign, and thus the main contribution to the jacking phenomenon is the inequality between the tyre forces. For suspensions in which the wheel camber is sensitive to bounce then this effect will lead to further inequality in lateral tyre forces. The swing axle suspension is notorious for this behaviour.

4.19 COMMENT

This chapter describes a means of specifying the behaviour of a suspension by considering the changes in suspension parameters induced by movements of the vehicle body and steering system, the kinematic characteristics, and the elastic deflections caused by forces in the plane of the road, the compliance characteristics. This allows the suspension to be completely characterised in its quasi-static modes. With the addition of damping curves for the shock absorber elements these characteristics provide all the data required in the study of the rigid body modes of vehicle dynamics.

A description of some suspension types is given as a way of familiarising the reader with the physical layout of the systems and the forces that the suspension is required to transmit from the road to the vehicle body.

Some basic two dimensional kinematic analyses then show how the detailed layout of the mechanism controls the attitude of the tyre on the road and hence affects the generation of the forces which guide and control a vehicle. It is important to note that the concept of the roll centre of a suspension is a kinematic facility for the purpose of graphic analysis, it does not imply that in a dynamic situation the vehicle will roll around an axis joining the roll centres. The suspension derivative analysis provides a more general description of the behaviour of a suspension than the roll centre method since it can be used when the body has attained a roll angle when the analysis will give different values of $dy'/d\phi$ for left and right wheels.

With the vehicle body in an upright position the value of $dy'/d\phi$ is equivalent to the vertical distance of the roll centre from the origin taken for the suspension derivative analysis.

These static analyses are capable of producing kinematic data for vehicle handling models.

Deflection of the suspension from the kinematic state occurs as a result of lateral and longitudinal forces in the plane of the road, the majority of this deflection, known as suspension compliance, is due to the mounting bushes.

For the higher frequency modes in which noise is transmitted through the suspension into the body it is necessary to have a dynamic analysis which considers the mass and inertias of each element of the suspension together with the stiffnesses of the bushings. There is no intrinsic difficulty in developing the analysis and a number of excellent mechanism programs are available which can readily be adapted for this purpose.

REFERENCES

(1) F.D.HALES (1965) Theoretical analysis of thr lateral properties of suspension systems, *Proc. Instn mech. Engrs*, **179**, Part 2A, 73-97.

(2) J.R.ELLIS (1965) Introduction to the dynamic properties of vehicle suspensions, *Proc. Instn mech. Engrs*, **179**, Part 2A, 98-112.

(3) J.R.ELLIS (1968) *A study of suspension mechanisms*, ASAE Report No.5, Cranfield Institute of Technology, Cranfield, Bedfordshire.

(4) D.M.BUTLER and J.R.ELLIS (1972) Analysis and measurement of relative movements between the wheels of a vehicle and the road surface, *Proc. Instn mech. Engrs*, **186**, 793-806.

(5) J.R.ELLIS, W.R.GARROTT, *et al.* (1986) *Measurement of the suspension characteristics of cars and light trucks*, H00349, American Society of Mechanical Engineers, New York.

CHAPTER FIVE

The Rigid Vehicle Roll Mode

5.1 INTRODUCTION

When a roll motion is added to the vehicle handling model it is seen that, although with the model derived there is no inertial coupling between the sideslip-yaw modes and the roll mode, some subtle changes in the vehicle characteristics may occur. Roll induced steering is possible at both the front and rear axle. There is a tranfer of normal force between the left and right side tyres during a turn, the relative roll stiffnesses of the suspensions will affect the proportion of that load tranfer carried by the front and rear axles.

Trim analysis provides a means of determining the responses to steering and wind gusts when the vehicle is in a state of non zero lateral acceleration. The method depends upon the fact that the equations of motion derived earlier in the chapter are not limited to small perturbations about the straight running condition but may be applied, with suitable tyre data, to motion about any steady state. The moment method examines the relation between the lateral force and yawing moment when a vehicle is treated as though it were constrained in an artificial environment in which it may be held at a sideslip angle to the road and the steer angle varied between extreme limits. The zero yawing moment condition indicates equilibrium.

A note on the aerodynamic characteristics of cars gives some examples of measured data and shows the steps required to transform the results from the `usual aerodynamic axes to the vehicle axes.

It is always necessary to qualify the amplitude, frequency and damping values obtained from this study with the fact that a car is required to operate on a roadway under the control of a driver. Although the subject of the interaction between car, driver and the road environment is complex and not well understood it appears that a driver can adapt, in time, to most vehicles but is confounded when the characteristics change without warning. Some examples of these changes are demonstrated in various parts of this chapter.

5.2 THE CAR WITH ROLL

The effects of roll motion on the handling characteristics of a car are due to the following changes in lateral tyre forces and aligning moments caused by roll.

(1) Change in normal force at tyre/road interfaces.
(2) Steering of the road wheels with roll displacement.
(3) The component of the lateral velocity at each interface due to tyre scrub.
(4) Camber of the wheels with roll displacement.

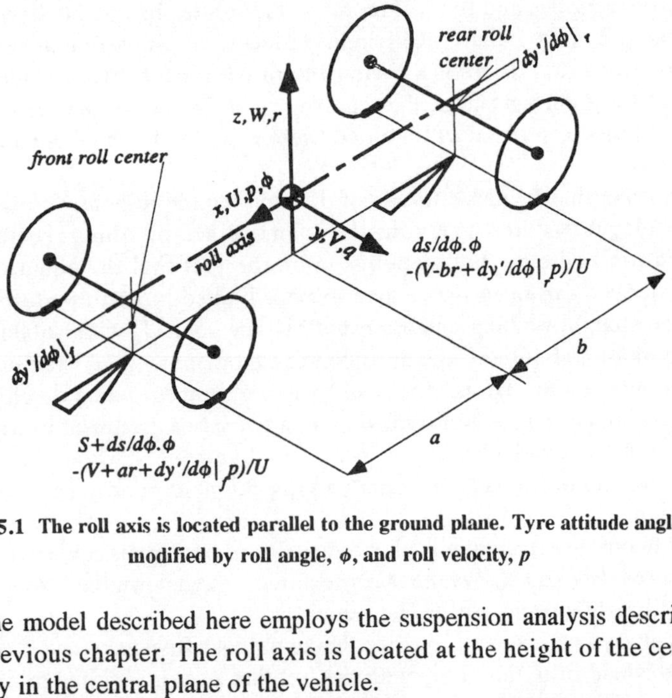

Fig.5.1 The roll axis is located parallel to the ground plane. Tyre attitude angles are modified by roll angle, ϕ, and roll velocity, p

The model described here employs the suspension analysis described in the previous chapter. The roll axis is located at the height of the centre of gravity in the central plane of the vehicle.

The roll moment of inertia includes terms which describe inertial characteristics of the road wheels due to roll.

The tyre contact surfaces are displaced laterally by roll. When the velocity of this lateral (scrubbing) motion is combined with the forward speed of the car an additional slip angle is generated $(dy'/d\phi)p/U$. The road wheels may also be steered in proportion to the roll angle, $ds/d\phi \cdot \phi$, and a camber angle will also be present, $d\phi'/d\phi \cdot \phi$.

Thus the equations of motion are for the special case where the centre of mass is located at the origin with bounce, W, and pitch, q, velocities ignored:

$$\Sigma X = m(\dot{U} - Vr)\ ; \quad \Sigma Y = m(\dot{V} + Ur)$$
$$\Sigma N = I_z \dot{r}\ ; \quad \Sigma L = I_x \dot{p} \tag{5.1}$$

5.3 THE EFFECTIVE ROLL INERTIA OF A RIGID VEHICLE

When roll motion is included as a degree of freedom both the vehicle body and the suspension units move, the relative motions are described by the suspension derivatives and thus the body and suspensions may be considered as geared together with the result that a single inertial term may be used for the roll of the system.

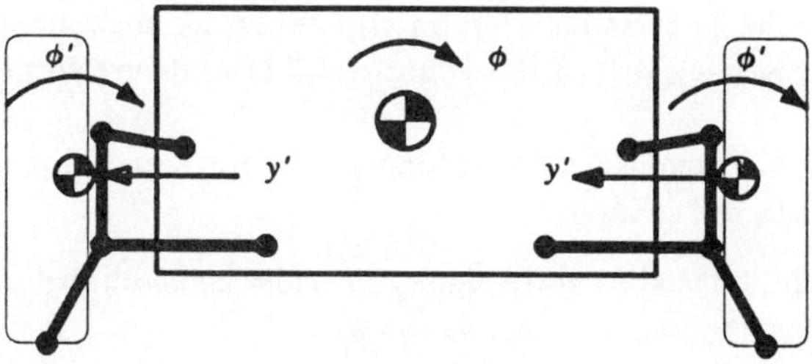

Fig.5.2 As the vehicle body rolls the wheels camber and move laterally under suspension control

In general tyre lateral force data are not available for conditions under which the road wheel resonates due to road irregularity and while the wheels are in contact with a relatively smooth surface the motion of the wheel is determined by movement of the vehicle body as described in the chapter on suspension design and analysis. Thus the inertial behaviour of the wheels may be described in terms of the bounce, roll and pitch of the body. For the case in which roll is the only mode of motion Fig.5.2. shows that the motion of the road wheels in response to roll of the car is constrained by the kinematic design of the suspensions. Similar considerations may be applied to the wheel movements due to pitch and bounce.

Consider the kinetic energy of the system due to roll.

$$2T = Ip^2 + (m'V'^2 + I'p'^2)$$

but

$$V' = dy'/d\phi \,.\, p$$

and

$$p' = d\phi'/d\phi \,.\, p$$

These suspension derivatives refer to the centres of mass of the suspensions. Thus the effective inertia in roll of the whole car is:

$$I_x = I + \Sigma m'(dy'/d\phi)^2 + \Sigma I'(d\phi'/d\phi)^2 \qquad (5.2)$$

5.4 TYRE SLIP ANGLE INDUCED BY ROLL VELOCITY

Once the vehicle analysis is released from the specification of a roll axis as the line joining the front and rear instantaneous centres, which effectively restrains the wheels from developing slip angles as the vehicle rolls, the interactions between roll of the vehicle and lateral tyre forces becomes clear.

There are two conditions connecting lateral force and aligning moment from the tyres to roll motion.

(1) Tyre scrub induced by roll velocity provides an additional term in the front and rear slip angles, $dy'/d\phi \,.\, p/U$.
(2) In many cases the suspensions of a car are designed to cause the wheels to steer as the body rolls, $ds/d\phi \,.\, \phi$.

Thus roll may change both the steer and slip angles of the tyres, a point illustrated in Fig.5.1.

$$\text{Front:} \quad \alpha_f = s + ds/d\phi_1\phi - (V + ar + dy/d\phi|_1 p)/U$$
$$\text{Rear:} \quad \alpha_r = ds/d\phi_2\phi - (V - br + dy/d\phi|_2 p)/U \qquad (5.3)$$

Equations (5.3) give the attitude angles from which the lateral force and aligning moment may be calculated.

Note that with the axis system located along the principal axes of the car there is no inertial coupling between the yaw - sideslip modes and the roll mode and the coupling between modes is confined to the modified tyre steer, slip, and camber.

5.5 LATERAL TYRE FORCE AND ROLL

Lateral tyre forces contribute to the roll motion of the car because roll of the car causes a lateral or scrub movement of the tyres on the road, when taken in conjunction with the lateral force this provides a work function which appears as a disturbance to the roll equation. Figure 5.3 shows this effect.

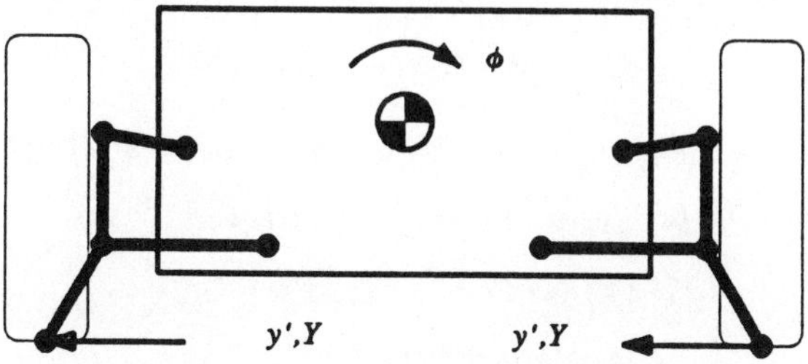

Fig.5.3 Work is done as the tyres scrub during roll

$$\text{Work} = \Sigma Y y'$$

$$\text{Work} = \Sigma(Y \,.\, dy'/d\phi)\phi \qquad (5.4)$$

5.6 EQUATIONS OF MOTION

The equations of motion may now be written in terms of the tyre characteristics, for the linear case. The following notation is introduced to make the equations more compact.

Tyre camber force coefficient: $Q = dY/d\phi$

Roll steer coefficient: $e = ds/d\phi$

Tyre scrub with roll: $g = dy/d\phi$

Wheel camber with roll: $q = d\phi'/d\phi$

The forces and moments generated by the tyres are given below.

$$\begin{aligned} Y_1 &= C_1 \{s+e_1\phi - (V+ar+g_1 p)/U\} + q_1 Q_1 \phi \\ N_1 &= \hat{T}_1 \{s+e_1\phi - (V+ar+g_1 p)/U\} \\ Y_2 &= C_2 \{e_2\phi - (V-br+g_2 p)/U\} + q_2 Q_2 \phi \\ N_2 &= \hat{T}_2 \{e_2\phi - (V-br+g_2 p)/U\} \end{aligned} \qquad (5.5)$$

Thus the equations of motion at constant speed, U, become:

$$
\begin{aligned}
m(\dot{V}+Ur) &= Y_1 + Y_2 \\
I_z\dot{r} &= aY_1 - bY_2 + N_1 + N_2 \\
I_x\dot{p} + C_\phi p + K_\phi \phi &= g_1 Y_1 + g_2 Y_2
\end{aligned}
\tag{5.6}
$$

In terms of the tyre attitude angles:

$$
\begin{aligned}
m(\dot{V}+Ur) &= C_1 \{s+e_1\phi - (V+ar+g_1 p)/U\} + q_1 Q_1 \phi \\
&\quad + C_2 (e_2\phi - (V-br+g_2 p)/U) + q_2 Q_2 \phi \\
I_z\dot{r} &= a \cdot [C_1 \{s+e_1\phi - (V+ar+g_1 p)/U\} + q_1 Q_1 \phi] \\
&\quad - b \cdot [C_2 \{e_2\phi - (V-br+g_2 p)/U\} + q_2 Q_2 \phi] \\
&\quad + \hat{T}_1 \{s+e_1\phi - (V+ar+g_1 p)/U\} \\
&\quad + \hat{T}_2 \{e_2\phi - (V-br+g_2 p)/U\} \\
I_x\dot{p} + C_\phi p + K_\phi \phi & \\
&= g_1 \cdot [C_1 \{s+e_1\phi - (V+ar+g_1 p)/U\} + q_1 Q_1 \phi] \\
&\quad + g_2 \cdot [C_2 \{e_2\phi - (V-br+g_2 p)/U\} + q_2 Q_2 \phi]
\end{aligned}
\tag{5.7}
$$

The separation of the variables shows how each affects the situation.

$$
\begin{aligned}
m(\dot{V}+Ur) &= C_1 s - (C_1+C_2)V/U - (aC_1-bC_2)r/U \\
&\quad - (g_1 C_1+g_2 C_2)p/U \\
&\quad + (e_1 C_1+e_2 C_2+q_1 Q_1+q_2 Q_2)\phi \\
I_z\dot{r} &= (aC_1+\hat{T}_1)s - (aC_1-bC_2+\hat{T}_1+\hat{T}_2)V/U \\
&\quad - (a^2C_1+b^2C_2+a\hat{T}_1-b\hat{T}_2)r/U \\
&\quad - (ag_1 C_1-bg_2 C_2+g_1\hat{T}_1+g_2\hat{T}_2)p/U \\
&\quad + (ae_1 C_1-be_2 C_2+e_1\hat{T}_1+e_2\hat{T}_2+aq_1 Q_1-bq_2 Q_2)\phi \\
I_x\dot{p} + C_\phi p + K_\phi \phi & \\
&= g_1 C_1 s - (g_1 C_1+g_2 C_2)V/U - (ag_1 C_1-bg_2 C_2)r/U \\
&\quad - (g_1^2 C_1+g_2^2 C_2)p/U \\
&\quad + (e_1 g_1 C_1+e_2 g_2 C_2+g_1 q_1 Q_1+g_2 q_2 Q_2)\phi
\end{aligned}
\tag{5.8}
$$

5.6 DERIVATIVE NOTATION

The coefficients from the two degree of freedom analysis are now augmented to account for roll.

$$Y_v = (C_1+C_2)/U$$
$$Y_r = (aC_1-bC_2)/U$$
$$Y_p = (g_1 C_1+g_2 C_2)/U$$
$$Y_\phi = -(e_1 C_1+e_2 C_2 + q_1 Q_1+q_2 Q_2)$$
$$Y_s = C_1$$
$$N_v = (aC_1 -bC_2+\hat{T}_1+\hat{T}_2)/U$$
$$N_r = (a^2C_1+b^2C_2 + a\hat{T}_1 -b\hat{T}_2)/U$$
$$N_p = (ag_1 C_1 -bg_2 C_2 + g_1\hat{T}_1+g_2 \hat{T}_2)/U \quad (5.9)$$
$$N_\phi = -(ae_1 C_1 -be_2 C_2 + e_1 \hat{T}_1+e_2 \hat{T}_2 + aq_1 Q_1 -bq_2 Q_2)$$
$$N_s = aC_1 + \hat{T}_1$$
$$L_v = (g_1 C_1 + g_2 C_2)/U$$
$$L_r = (ag_1 C_1 - bg_2 C_2)/U$$
$$L_p = C_\phi + (g_1^2 C_1+g_2^2 C_2)/U$$
$$L_\phi = K_\phi - (e_1 g_1 C_1+e_2 g_2 C_2 + q_1 g_1 \hat{C}_1+q_2 g_2 \hat{C}_2)$$
$$L_s = g_1 C_1$$

The matrix form of the equations demonstrate the nature of the interactions between the roll and the yaw-sideslip modes.

$$\begin{bmatrix} mD+Y_v & mU+Y_r & Y_pD+Y_\phi \\ N_v & I_zD+N_r & N_pD+N_\phi \\ L_v & L_r & I_xD^2+L_pD+L_\phi \end{bmatrix} \begin{bmatrix} v \\ r \\ \phi \end{bmatrix} = \begin{bmatrix} Y_s \\ N_s \\ L_s \end{bmatrix} s \quad (5.10)$$

The transient characteristics are mainly defined by the leading diagonal of the matrix. In this case the yaw and sideslip modes are similar to those of the two degree of freedom model. It appears that the yaw and sideslip responses are not greatly affected by the roll mode. However it will be seen that there is some feed back from roll due to the changes in the slip angles caused by tyre scrub induced by roll velocity.

The roll damping contains terms dependent upon the square of the tyre scrub with roll while the 'spring term' is dependent upon roll steer, camber force and tyre scrub as well as the suspension roll stiffness. Thus the roll motion is influenced by the sideslip/yaw modes and the damping is reduced since values of $dy/d\phi$ are almost always negative.

5.7 STEADY STATE RESPONSES TO STEERING FOR THE THREE-DEGREE-OF-FREEDOM MODEL

The steady state responses to steering are obtained when all the derivative terms are zero.

$$\begin{bmatrix} Y_v & mU+Y_r & Y_\phi \\ N_v & N_r & N_\phi \\ L_v & L_r & L_\phi \end{bmatrix} \begin{bmatrix} v \\ r \\ \phi \end{bmatrix} = \begin{bmatrix} Y_s \\ N_s \\ L_s \end{bmatrix} s \tag{5.11}$$

The responses are given below in determinant form.

$$\beta/s\big|_{ss} = \frac{\begin{vmatrix} Y_s & mU+Y_r & Y_\phi \\ N_s & N_r & N_\phi \\ L_s & L_r & L_\phi \end{vmatrix}}{\begin{vmatrix} Y_v & mU+Y_r & Y_\phi \\ N_v & N_r & N_\phi \\ L_v & L_r & L_\phi \end{vmatrix} U}$$

$$r/s\big|_{ss} = \frac{\begin{vmatrix} Y_v & Y_s & Y_\phi \\ N_v & N_s & N_\phi \\ L_v & L_s & L_\phi \end{vmatrix}}{\begin{vmatrix} Y_v & mU+Y_r & Y_\phi \\ N_v & N_r & N_\phi \\ L_v & L_r & L_\phi \end{vmatrix}} \tag{5.12}$$

$$\phi/s\Big|_{ss} = \frac{\begin{vmatrix} Y_v & mU+Y_r & Y_s \\ N_v & N_r & N_s \\ L_v & L_r & L_s \end{vmatrix}}{\begin{vmatrix} Y_v & mU+Y_r & Y_\phi \\ N_v & N_r & N_\phi \\ L_v & L_r & L_\phi \end{vmatrix}}$$

5.8 STEADY STATE TURNING ON A FIXED RADIUS

The steady state lateral acceleration responses are obtained from the matrix by deleting all derivative terms and transposing the yaw velocity and steer columns.

$$\begin{bmatrix} Y_v & Y_\phi & -Y_s \\ N_v & N_\phi & -N_s \\ L_v & L_\phi & -L_s \end{bmatrix} \begin{bmatrix} v \\ s \\ \phi \end{bmatrix} = \begin{bmatrix} -(mU+Y_r) \\ -N_r \\ -L_r \end{bmatrix} (U^2/R) \tag{5.13}$$

$$\beta\Big|_{ss} = \frac{\begin{vmatrix} -(mU+Y_r) & Y_\phi & -Y_s \\ -N_r & N_\phi & -N_s \\ -L_r & L_\phi & -L_s \end{vmatrix}}{\begin{vmatrix} Y_v & Y_\phi & -Y_s \\ N_v & N_\phi & -N_s \\ L_v & L_\phi & -L_s \end{vmatrix}} (U^2/R)$$

$$\phi\Big|_{ss} = \frac{\begin{vmatrix} Y_v & -(mU+Y_r) & -Y_s \\ N_v & -N_r & -N_s \\ L_v & -L_r & -L_s \end{vmatrix}}{\begin{vmatrix} Y_v & Y_\phi & -Y_s \\ N_v & N_\phi & -N_s \\ L_v & L_\phi & -L_s \end{vmatrix}} (U^2/R) \tag{5.14}$$

$$s|_{ss} = \frac{\begin{vmatrix} Y_v & Y_\phi & -(mU+Y_r) \\ N_v & N_\phi & -N_r \\ L_v & L_\phi & -L_r \end{vmatrix}}{\begin{vmatrix} Y_v & Y_\phi & -Y_s \\ N_v & N_\phi & -N_s \\ L_v & L_\phi & -L_s \end{vmatrix}} (U^2/R)$$

5.9 STEADY STATE SIDESLIP/YAW BALANCE

The effects of introducing a further variable, roll angle, into the equations is not only to change the values of the variables in a particular condition but also to alter the balance between the variables. In this model the relative values of sideslip and yawing velocity are affected thus a relation between these variables should be established. The steady state relation is given below.

$$\beta/r|_{ss} = \frac{\begin{vmatrix} Y_s & mU+Y_r & Y_\phi \\ N_s & N_r & N_\phi \\ L_s & L_r & L_\phi \end{vmatrix}}{\begin{vmatrix} Y_v & Y_s & Y_\phi \\ N_v & N_s & N_\phi \\ L_v & L_s & L_\phi \end{vmatrix} U} \tag{5.15}$$

5.10 THE EFFECTS OF ROLL ON HANDLING CHARACTERISTICS

Roll of the vehicle may affect:

(1) Lateral force of the road wheels by cambering the wheels ($d\phi'/d\phi$).
(2) The position of the roll centres ($dy'/d\phi$) may modify the handling responses by changing the normal forces on the tyres.
(3) F&M due to tyre scrub ($dy'/d\phi$).
(4) F&M by roll steer ($ds/d\phi$).

Thus the roll characteristics of the suspensions provide a subtle means of modifying handling responses. This section examines some of the possibilities using the steady state and transient equations, the study is not exhaustive.

If Y_ϕ and N_ϕ are zero, which is true if there is no roll steer or camber, then the characteristic equation is similar to that for the two degree of freedom car except for the roll stiffness factor.

Camber thrust

In almost all suspensions the road wheels either remain upright or take up a camber angle less than the roll angle and in the same sense.

Under steady state conditions camber at the front wheels will generate understeer, while at the rear wheels camber will induce oversteer. During transients the lateral forces are decreased in phase with the roll angle. Many vehicles operate with the rear wheels remaining upright while the front wheels take up a camber angle and this type of vehicle will have increasing understeer with roll angle.

Tyre scrub with roll

The suspension characteristic $dy'/d\phi$ is the tyre lateral scrub rate with roll. It will be noted that it is also a statement of the *'roll centre* for small roll angles. For most practical suspensions the roll centre lies between ground level <parallel arms> and the height of the line between the wheel centres <beam axle>. In the steady state the terms $g\ (dy'/d\phi)$ do not appear in the sideslip and yaw equations (5.8) and thus tyre scrub, or the roll centre position does not affect sideslip *(β)* or yaw velocity *(r)*. Roll angle increases as the roll centres become lower. For example on a vehicle with the rear roll centre at the wheel centre height $[dy'/d\phi|_r = -0.4$, cg at 0.7m] lowering the front roll centre from $[dy'/d\phi|_f = -0.4]$, the wheel centre to $[dy'/d\phi|_f = -0.8]$, 0.1m below the road surface, increases the roll angle by 40 percent for the model used here.

Transient mode coupling of sideslip/yaw to roll is through the roll velocity terms Y_p and N_p respectively. The simulations shown here represent the effects of a single sinusoid of seven second period. tyre characteristics are assumed constant and not modified due to load tranfer during the test. This is in order to isolate the effects of varying $dy'/d\phi$ at the front axle without introducing tyre load transfer effects which vary with different tyre designs. Note that the lowest roll centre at the front end, 0.1m below the road surface, gives the lowest sideslip, yaw velocity and lateral acceleration with the largest roll angle. There is a small decrease in the time lags for sideslip and yaw velocity as the front roll centre moves upward, this is shown by the earlier times for maximum values to be attained in each half cycle and also at the zero crossing points.

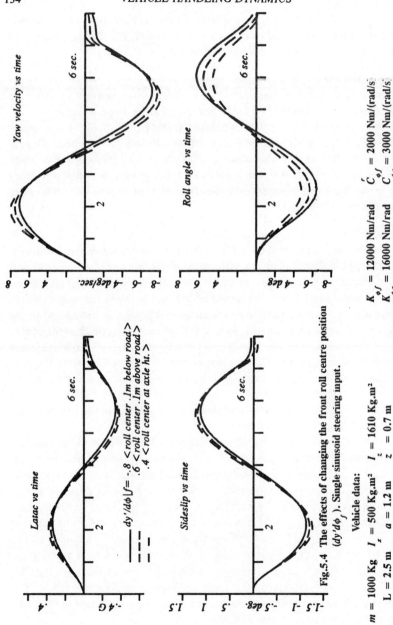

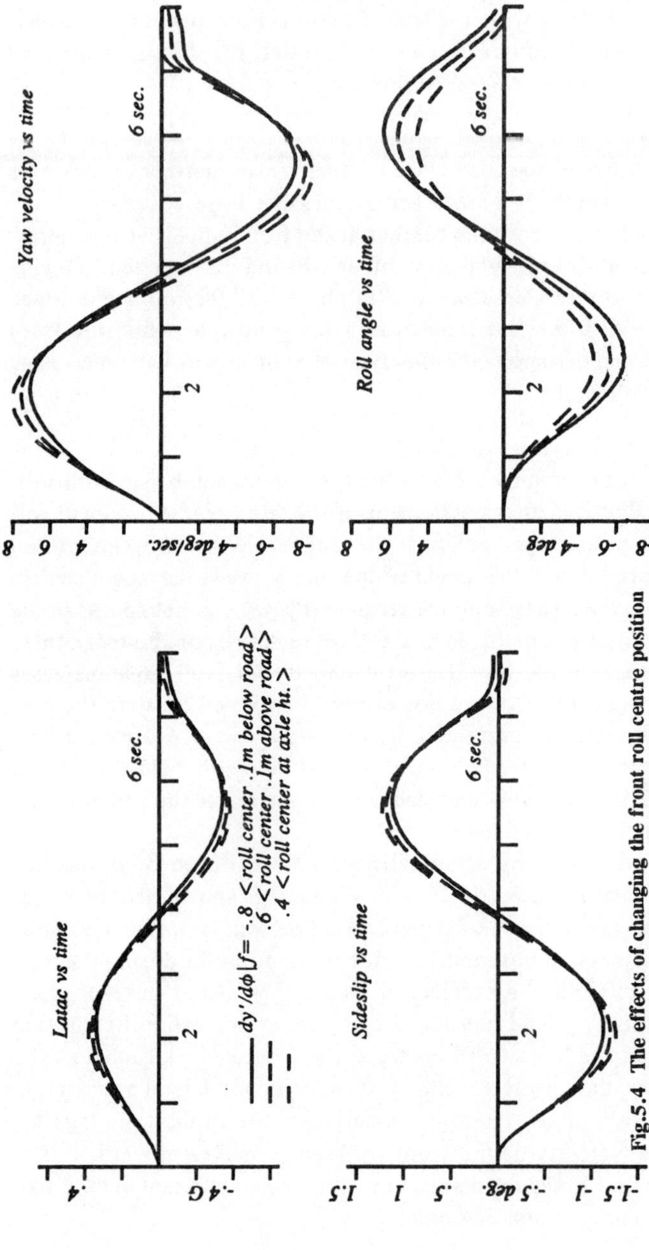

Fig.5.4 The effects of changing the front roll centre position ($dy'/d\phi_f$). Single sinusoid steering input.

Vehicle data:

m = 1000 Kg $\quad I_x$ = 500 Kg.m² $\quad I_z$ = 1610 Kg.m² $\quad K_{\phi f}$ = 12000 Nm/rad $\quad C_{\phi f}$ = 2000 Nm/(rad/s)

L = 2.5 m $\quad a$ = 1.2 m $\quad z_{cg}$ = 0.7 m $\quad K_{\phi r}$ = 16000 Nm/rad $\quad C_{\phi r}$ = 3000 Nm/(rad/s)

$dy'/d\phi$ = $\quad dY/ds$ = 32 KN/rad $\quad dN/ds$ = -600 Nm/rad $\quad dY/d\phi$ = 1.2 KN/rad

Roll steer

The term Y_ϕ contains values for the roll steer coefficient multiplied with the tyre lateral force coefficient $<eC = ds/d\phi.C>$. Positive roll steer at the front of the car increases the steady state understeer, while negative roll steer at the rear has a similar steady state effect. In a fixed radius turn the steering handwheel angle caused by a given positive value of $ds/d\phi$ at the front is nearly equal to that from a negative $ds/d\phi$ at the rear.

For any given condition of speed and radius of turn the lateral tyre forces are fixed, thus the total attitude angles are unchanged and the effect of any change in the steer angle introduced by roll, or by any other means, is simply to modify the values of the other variables. When roll steer is used at the front axle the sideslip angle of the car is not changed and it can be deduced that front roll steer is, in the steady state, a direct substitute for steering applied through the handwheel. This is not the case for the rear axle where the sideslip angle is modified by roll steer.

The transient responses to a single sinusoid steer input are shown in Figs 5.5 and 5.6. For roll steer at the front suspension the lateral acceleration plots show a negative $ds/d\phi$ reducing the peak lateral acceleration and the time lags, an indication that understeer is present. Similar conclusions can be reached from examination of the sideslip and yaw velocity plots.

Rear roll steer produces the opposite effect to front roll steer for lateral acceleration and yaw velocity; however, the sideslip angle is greatly changed in the presence of rear roll steer. This may affect the driver's perception of the vehicle.

The steering wheel input is 15 degrees thus the maximum steer angle at the road wheels is less than 1 degree. The additional steer from roll amounts to around 0.08 degree. A change of this magnitude in the suspension characteristic is difficult to detect and may well be less than the production tolerances.

Load transfer between the road wheels

As a car moves on a curved path the lateral inertial force acting at the centre of mass of the car gives rise to an overturning moment. This moment increases the normal force on the outer wheels and decreases the normal force on the inner wheels. Redistibution of normal force causes the lateral forces generated by the tyres moving at given attitude angles to change in a non-linear way and thus the balance of the car may change. These effects are frequently used; for example in rear drive cars an anti-roll or 'sway' bar may connect the two front suspensions so as to act in the roll

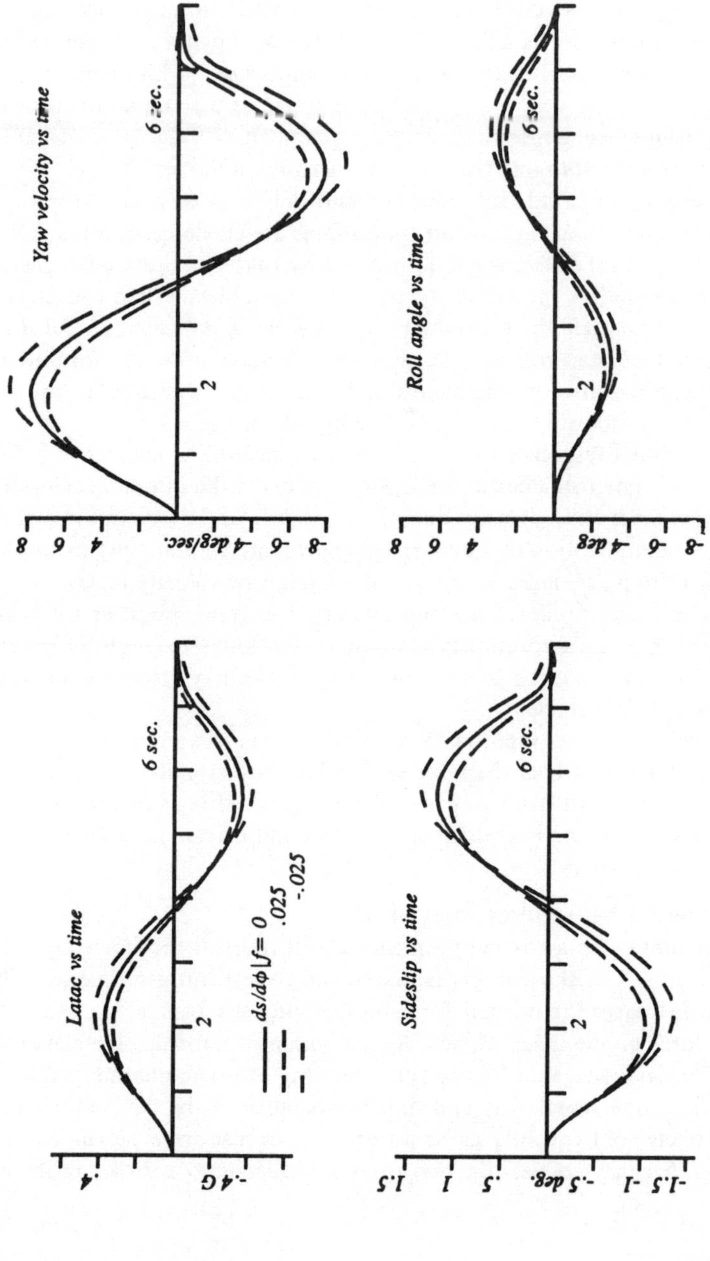

Fig.5.5 Roll steer at the front suspension

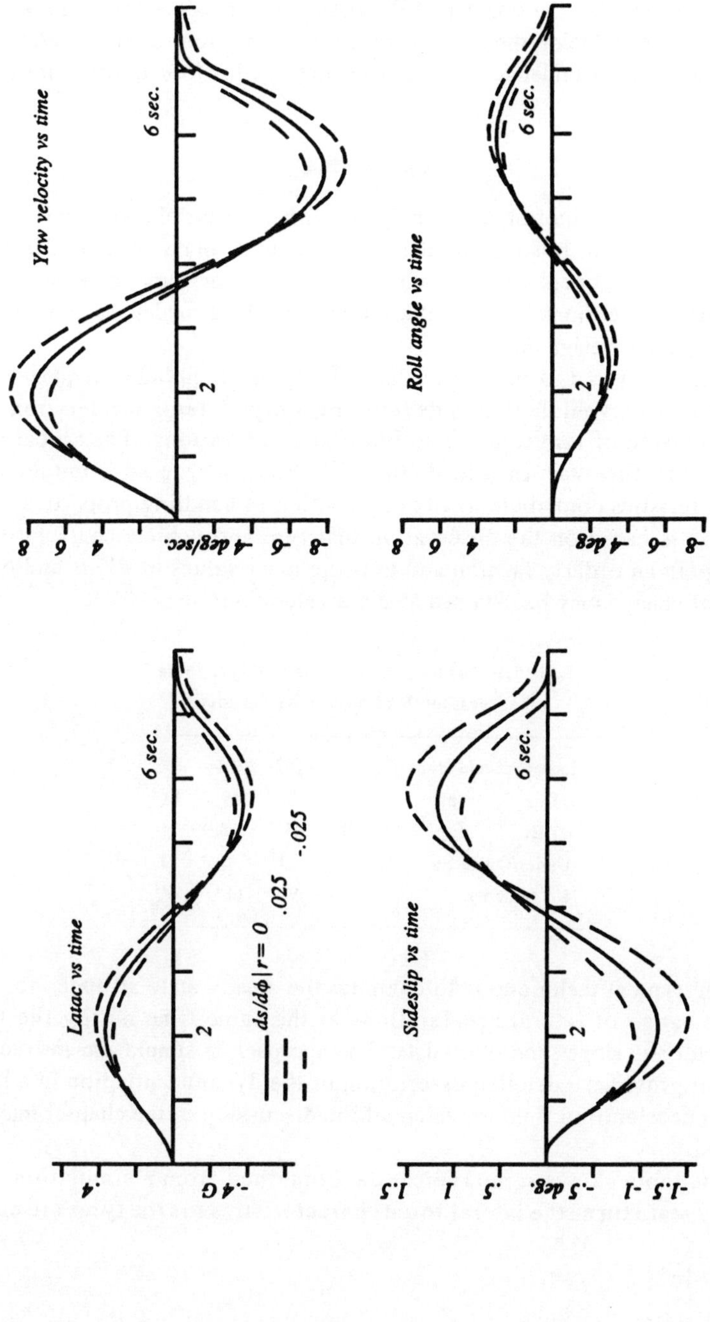

Fig.5.6 Roll steer at the rear suspension

mode and increase the transfer of normal force across the front wheels with the object of reducing the total lateral force at the front during steering and thus inducing an understeer effect. Calculations for load transfer are given in section 4.16.

5.11 TRIM ANALYSIS

The equations of motion previously developed in this chapter apply at any vehicle condition, braking, accelerating or turning alone or in combination. A number of models which contain non linear functions representing the tyres and suspensions have been developed but these models have not added notably to the design data base.

Trim analysis is a method of determining the amplitude, frequency and damping of a vehicle that is developing a high lateral acceleration or a combination of longitudinal and lateral accelerations. The object is to illustrate the way in which the changes in tyre and suspension characteristics contribute to any degradation in handling properties. The method is based on the observation that tyre and vehicle characteristics change in an orderly fashion and thus the local values of dY/ds and other rates of change may be obtained at any acceleration.

Table 5.1 The local slopes of lateral tyre force at several levels of lateral acceleration

Lateral acceleration (G)	0	0.2	0.4	0.6
dY/ds_f (kN/rad)	28	24	19	13
dY/ds_r	32	25.8	18.9	10.2
Under/Oversteer	U	U	U	O
U_c (mile/h)	84	90	114	60

The typical technique is to compute the steady state attitudes for the whole range of lateral accelerations at the same time noting the tyre characteristic slopes and other data. The approach is simple and instructive since it provides a detailed description of the dynamic situation in a high lateral acceleration field to which all the discussion of the chapter may be applied.

The table shows some typical data obtained from a simulation of a steady state turn, the lateral force characteristics for the tyres are cubic functions and the local values of the slope, dY/ds, are given for an understeer, rear drive car at various lateral accelerations. From the data the local under/oversteer condition is established. In this example the characteristic speed increases with lateral acceleration until at the highest lateral acceleration the car becomes oversteer. Simulations may be set up with the tyre data given in the table to demonstrate the changing nature of the transient responses in the non zero acceleration field.

5.12 AERODYNAMICS OF THE CAR

As the cost of fuel rises more attention is being paid to aerodynamic resistance to motion. This resistance comes not only from the drag but also from the work lost through the tyres. The car does not operate on windless roads and for the majority of its life it is subject to sidewinds. These sidewinds generate not only drag but also lateral force and yawing moment. Lateral force may be balanced by giving the car a sideslip angle. Steering the front wheels provides a balance to the yawing moment. The lateral tyre forces required dissipate energy and thus consume fuel.

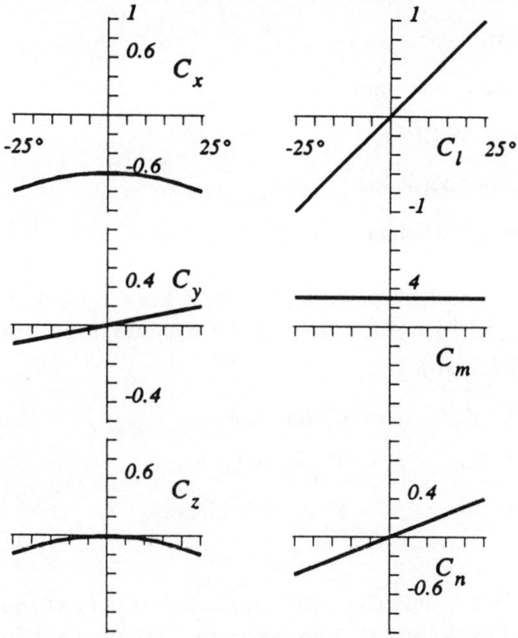

Fig.5.7 Typical aerodynamic force & moment data for a 'three box' shape

Aerodynamic values are usually expressed as coefficients, the forces and moments acting on a vehicle are dependent upon the pressure due to flow of the air, an area and in the case of the moment a representative length. In vehicle technology the area is the maximium cross section of the vehicle. It is convenient to remember the 'static head' as a number.

$$q = \tfrac{1}{2}\sigma U^2 \qquad q = (U/1.28)^2 \tag{5.16}$$

where U = m/s

Figure 5.7. shows some typical aerodynamic coefficients for a car from which it is seen that these are dependent upon the angle between the x axis of the car and the relative wind. Either these coefficients are written as polynomials or a look up table is used when calculating the effects of aerodynamic forces and moments on handling. Typical polynomials are shown below.

$$\begin{array}{ll} \text{Longitudinal coefficient:} & C_x = {}_0C_x + {}_2C_x a^2 \\ \text{Lateral coefficient:} & C_y = {}_1C_y a \\ \text{Normal force coefficient:} & C_z = {}_0C_z + {}_2C_z a^2 \\ \text{Roll moment coefficient:} & C_l = {}_1C_l a \\ \text{Pitch moment coefficient:} & C_m = {}_0C_m \\ \text{Yaw moment coefficient:} & C_n = {}_1C_n a \end{array} \tag{5.17}$$

The forces are calculated using the static head, q, and the vehicle cross section. A typical length is used in the computation of the moments as shown in the list below.

$$\begin{array}{ll} X = C_x Aq. & L = C_l Aq.\ \text{track} \\ Y = C_y Aq. & M = C_m Aq.\ \text{wheelbase} \\ Z = C_z Aq. & N = C_n Aq.\ \text{wheelbase} \end{array} \tag{5.18}$$

The origin for aerodynamic measurements is at the centre of the wheelbase and track, and it is necessary to carry out an axis transfer to the vehicle axis set. This is carried out as follows:

(a) measure forces and moments about the original axes;
(b) Transfer forces directly to the vehicle dynamics axes;
(c) calculate the moments around the dynamic axes.

Dynamic moment

= original moment + Forces · Separation of axes

The process is illustrated in Fig.5.7. Note that the moments and forces are first calculated and then transfered. Do not attempt to transfer coefficients.

The moments now become:

$$
\begin{aligned}
L &= L_a - Y\dot{z} \\
M &= M_a + Z\,(l/2-a) - X\dot{z} \\
N &= N_a - Y\,(l/2-a)
\end{aligned}
\tag{5.19}
$$

As the car moves on the road its attitude relative to the wind changes. Figure 5.8. shows a car with a heading angle ψ. The ambient wind velocity is expressed as components U_g and V_g parallel to the x and y axes, fixed in the road, from which the car heading angle is measured.

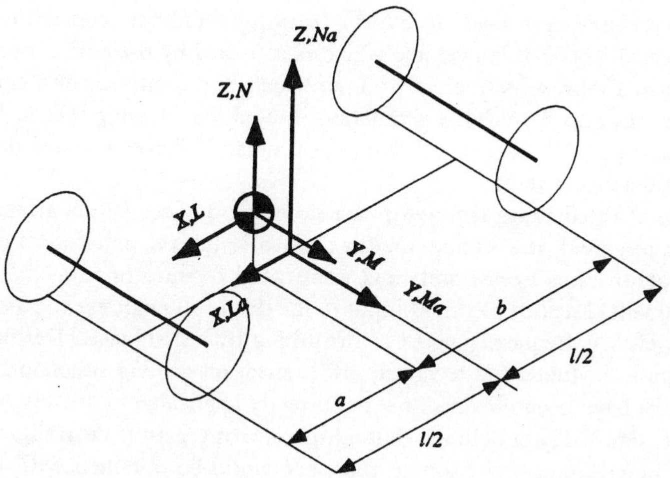

Fig.5.8 Aerodynamic data are measured from the centre of the wheelbase and then transferred to the vehicle dynamic axes

The heading angle of the car is obtained by open loop integration of yaw velocity. Aerodynamic forces and moments are calculated using an 'equivalent air speed' (EAS) which is the total velocity of the air relative to the car.

$$
\begin{aligned}
U_a &= U - U_g \cos\psi - V_g \sin\psi \\
V_a &= -V - U_g \sin\psi + V_g \cos\psi \\
\text{EAS} &= (U_a^2 + V_a^2)
\end{aligned}
\qquad (5.20)
$$

The angle of the airflow is the vector resultant of U_a and V_a

$$\tan\alpha = (V_a / U_a)$$

5.13 CAR AND DRIVER

Although a fairly rigorous description of the dynamics of the car has been developed there is no comparable model for the driver. A number of attempts have been made to build a 'driving simulator' consisting of a mathematical model of the vehicle which is activated by the human operator working in a representative environment such as the interior of a car and communicating to the vehicle simulation through the steering wheel, brake and throttle pedals. The output from the simulation drives a visual display to which the driver reacts.

Some of the driving simulators are complex devices which attempt to provide physical and visual feedback. Roll, lateral and longitudinal acceleration cues are sometimes generated by moving the driver's compartment. Lack of physical input to the driver in response to steering, braking, etc., is frequently noted as disturbing by test subjects. Despite the very significant funds and technical effort spent on driving simulators very few results have been obtained that are directly applicable to car design, for example, one study concluded that a high steering gain is desirable while another noted the a high yawing response could be obtained with radial tyres. On the other hand a simulator can be useful in a study of ways of alerting the driver to road conditions with road signs, radio messages, and other signals.

5.14 THE PATH CURVATURE

A vehicle moves on a road on a curved path which is described using the velocities of the centre of gravity. The general expression for curvature is given below.

$$1/R = d^2y/dx^2/\{1+(dy/dx)^2\}^{3/2} \tag{5.21}$$

When $x = f(t)$ and $y = g(t)$, the curvature may be written as

$$1/R = \frac{dy/dt.d^2x/dt^2 - dx/dt.d^2y/dt^2}{(dy/dt)^2\{1 + (dx/dt/dy/dt)^2\}^{3/2}} \tag{5.22}$$

Now

$$dx/dt = U\cos\psi - V\sin\psi$$

$$dy/dt = U\sin\psi + V\cos\psi$$

hence

$$d^2x/dt^2 = (\dot{U}-Vr)\cos\psi - (\dot{V}+Ur)\sin\psi$$

$$d^2y/dt^2 = (\dot{U}-Vr)\sin\psi + (\dot{V}+Ur)\cos\psi$$

By substituting these expressions in the curvature equation and writing $a(x) = \dot{U}-Vr$ and $a(y) = \dot{V}+Ur$, the following expression is obtained.

$$1/R = \{Ua(y) - Va(x)\}/(U^2 + V^2)^{3/2} \tag{5.23}$$

Let $\beta = V/U$, then

$$1/R = \{a(y) - \beta\, a(x)\}/(1+\beta^2)^{3/2}$$

Alternatively, equation (5.23) may be written

$$1/R = \{Ua(y) - V\,a(x)\}/V^3 \tag{5.24}$$

where

$$V = \sqrt{(U^2+V^2)}$$

Equation (5.24) shows that the path curvature is dependent upon both the lateral and longitudinal accelerations and consists of both steady state and transient parts. Since $V \approx U$ and the term $Va(x)$ is small when the car is moving at a steady forward speed the curvature for that condition may be approximated by a much simpler expression.

$$1/R = a(y)/U^2 \tag{5.25}$$

For the steady state turn the curvature is

$$1/R = r/V = r/U \tag{5.26}$$

With the approximations specified above the steering responses can be examined as changes in curvature. Two sets of data are needed: the yaw velocity response, from which the steady state component will be determined, and the lateral acceleration history for the total curvature.

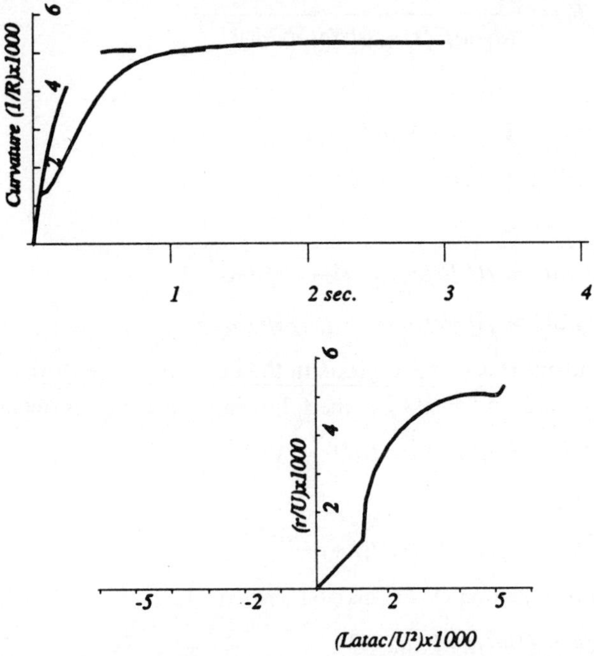

Fig.5.9 In a step steer test the steady state curvature component exceeds the total curvature until steady state is reached

The diagrams in Fig.5.9 show, in the lower graph, the total curvature as the vertical scale, and the apparent or steady state component derived from consideration of the yawing velocity as the horizontal axis. The right hand graphs are time histories of the total curvature (lateral acceleration/U^2) and steady state component (r/U), using the approximations derived for a constant speed car. The plots show that the apparent curvature of the path is always greater than the real curvature until a steady state is reached.

Thus it would appear that the driver is receiving two sets of information, the apparent or steady state curvature which is a visual stimulus and the true curvature which is felt as a lateral acceleration. In the case of a step or ramp input of steering the differences may not be important.

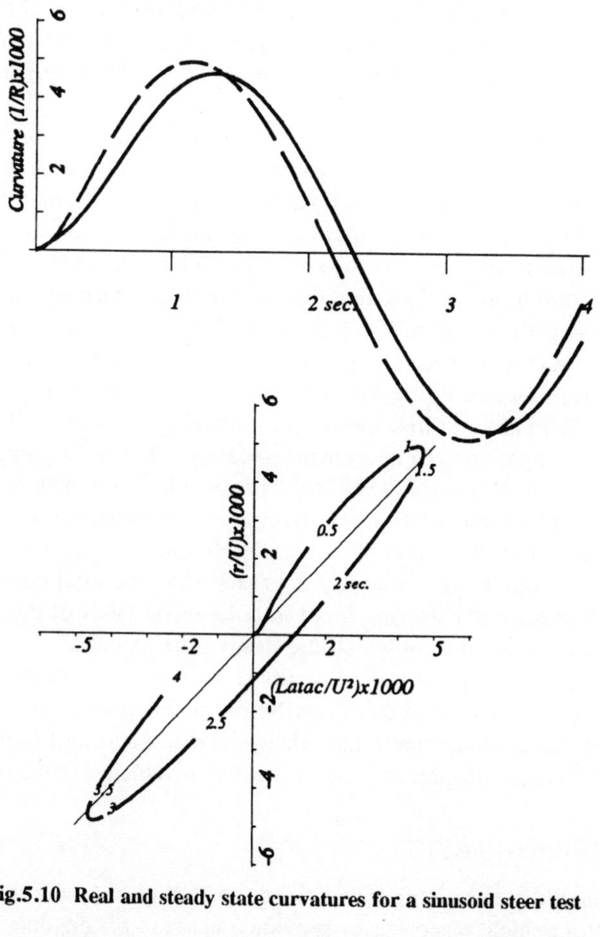

Fig.5.10 Real and steady state curvatures for a sinusoid steer test

When the curvature concept is applied to a single sinusoid steering input of four seconds time period, as shown in Fig.5.10 then the reults are more interesting. The first 1/4 cycle extends from time zero to 1.3 s. During this time the steady state or apparent curvature is greater than the actual curvature with the result that the car appears to be turning on a smaller radius than is actually the case. This is followed by the mid period of the motion during which the actual curvature exceeds the apparent value and both change from positive to negative. The final phase of the steering results in the apparent curvature again being larger than the real value.

In their original paper the authors provided an interpretation of the results obtained from the single sinusoid steering test which provided an insight into the interactions between car and driver. This is broadly paraphrased below.

Lateral acceleration and yaw velocity are perceived by the driver as cues to the vehicle responses to steering. In the initial phase of a manoeuver turning the steering wheel causes a yawing moment to be generated which the driver observes as a lateral motion of the horizon. This is followed by lateral acceleration as both tyres develop slip angles. Thus a driver demands a path curvature based on yaw rate that is an implied steady state response. However, the path curvature is a function of both steady state and transient conditions, and since there is a greater time lag between acceleration and steering than between yaw velocity and steering the driver receives conflicting signals from his visual and inertial perceptions. This is not important in single control movement situations, but for a complex steer programme, such as required for a lane change, the driver may be confused by the non-agreement of the two stimuli. Considering the sinusoidal response curvature plots it can be seen that in the first part of the process the component due to yaw velocity is greater than the total curvature. As the driver decreases the steering input into the second half of the sine wave then there is a period during which the steady state and total curvature have different polarity. A similar period occurs at the end of the steering cycle. It is suggested that these periods of conflicting polarity account for the well documented observations that a lane change is disconcerting to drivers. The lane change is also considered to be influential in vehicle evaluation.

5.15 HANDLING TESTS

A set of handling tests is now described. Although they do not establish standards for vehicle responses or provide a measure of 'goodness' they are interesting because they reflect the general atmosphere of vehicle handling tests within the international automotive community.

A wide range of variables is available for measurement and it is necessary to the equations of motion, existing measuring devices and the vehicle itself before testing starts. For example, the steered angles of the front wheels is used as an input to the basic equations but an examination of the steering mechanism soon shows the difficulty of locating a device such as a potentio- meter at the stub axle thus it is the steering handwheel angle and steering torque which are measured and the actual steer angle is estimated by calibrating these devices prior to the test. It has been shown that the inertia of the steering handwheel affects the free control mode of handling and thus care will be taken to ensure that the instrumented steering wheel has similar inertial characteristics to the original unit. Lateral velocity, or sideslip angle are difficult to measure and although instruments are available for this purpose they are complex and not usually included in a basic package.

The Table 5.2 gives a list of instrumentation which is widely available and easy to install, together with the power packs and invertors required. This table is compiled from references (5.2)-(5.5).

Table 5.2 A list of transducers frequently used for handling tests

Variable	*Range*	*Resolution*
Steering wheel angle	$\pm$ 360 degrees	$\pm$ 2 degrees
Lateral acceleration	$\pm$ 15 m/s^2	1% FS
Forward speed	0 - 50 m/s	1% FS
Yaw velocity	$\pm$ 50 degree/s	1% FS
Steering wheel torque	30 Nm	1% FS
Roll angle	$\pm$ 15 degrees	1% FS
Roll velocity	$\pm$ 50 degree/s	1% FS
Vertical acceleration	$\pm$ 60 m/s^2	1% FS
Longitudinal acceleration	$\pm$ 15 m/s^2	1% FS

Brake line pressure, brake temperature, brake pedal pressure, fuel flow and wheel to body movement are other variables that it may be deemed necessary to measure from time to time.

The tests consist of one steady state and a number of transient procedures. A list of basic transducers required is given together with their ranges and the recommended maximum errors of the transducer/recorder system. The first four items are essential components for a handling package while the others may be considered desirable. Most modern recorders are based on digital computers thus it is necessary to use analog to digital conversion, 12 bits are sufficient to meet the 1 percent FS specification for the system. A typical sampling rate of 50 Hz is satisfactory for handling experiments. The recorder should be capable of being programmed to give a few channels of high frequency sampling if it is desired to capture transients of brake operation with anti lock brakes. An integral part of the system is a 'quick look' facility which will provide a rapid check after each run that data has been captured.

The steady state circular test is designed to measure the steering wheel angle as a function of steady state lateral acceleration and to describe the understeer/oversteer for left and right turns. A plot of steer angle versus lateral acceleration (lateral acceleration/gravity constant) provides the ratio angle/G which defines understeer. The test is carried out on a plane surface or suitable road at various speeds and radii.

Transient responses can be examined in either the time or frequency domain. For the time domain the features of interest are the time lag, response time, overshoot and gain between the steering wheel input and the lateral accceleration and yaw velocity. These values have some correlation with the subjective aspects of car handling. Typical tests in the time domain are:

- step steering input;
- single sinusoid steering.

In the frequency domain there are four tests from which it is theoretically possible to compute the desired transfer functions:

- step steering;
- random steering;
- pulse steering;
- continuous sweep sinusoid steering.

In practice there is a general opinion that these procedures are not equal. One group of experimenters concludes that the step/ramp, single sinusoid and random steering each give unique results which are not directly comparable and thus all tests are required. Another author strongly supports the random steer input after a series of tests with step/ramp steering, the single sinusoid sweep steering and random steer.

The step/ramp steer is useful only for steady state responses and yields response times which do not uniquely define the vehicle.

Single sinusoid steering provides no frequency related data, but is favored by some manufacturers as a model for a lane change.

Random steer testing, carefully controlled, provides good data capable of discriminating between different vehicles, reliability can be confirmed by coherence computations. It was found that five runs each of thirteen seconds duration provides sufficient data for analysis.

Sweep steering with variable frequency has been suggested as a alternative to random steering, because it has been found that in practice the random steer tests give little input energy in the range 0 - 1 rad/s. It was noted that in practice the sweep steer does not contribute to the widening of the frequency range and restricts the energy in the frequency band below 3 rad/s.

Braking in a turn has also been used as a means of assessing vehicles and may be useful in defining limits of control under wet conditions.

Simulation of these test methods using linear models provides an interesting experience prior to actual testing.

5.16 COMMENT

When the roll mode is introduced as a further degree of freedom the model changes considerably because the tyre slip angles now include terms which use the suspension characteristic $dy'/d\phi$, the tyre scrub due to roll (which defines the 'roll centre') to change the transient tyre forces. Roll steer, $ds/d\phi$, which has both steady state and transient effects on the lateral forces generated by the tyres. Wheel camber with roll, $d\phi'/d\phi$, also modifies tyre lateral force.

Trim analysis, a simple method which shows the responses of a car including under/oversteer at every level of lateral and longitudinal acceleration was shown using data from the original steady state turn model.

Simulations have been employed as the basis for vehicle driving simulators in which a driver controls the model by working in a simulated environment. The results from these studies have been sparse. An interesting, but untried method of examining the performance of a driver proposes that the difference between the actual path curvature and the path curvature as perceived visually by the driver is responsible for the inability of the driver to perform well in situations such as a lane change.

The International Organisation for Standardisation has complied a number of standards and proposals for vehicle handling testing. These form the basis for the section on practical testing.

REFERENCES

(1) J.R.ELLIS (1973) *A six degree of freedom vehicle model*, 2nd IAVSD Symposium, Paris.

(2) ISO DIS 4138.2 *Steady state circular test*, International Organisation for Standardisation.

(3) ISO/TC22/SC9/N185 (1979) *Road vehicle transient response test procedure (step/ramp input)*, Draft proposal for an International Standard.

(4) ISO/TC22/SC9/N194 (1979) *Road vehicle transient response test procedure (random input)*, Draft proposal for an International Standard.

(5) ISO/TC22/SC9/N191 (1979) *Road vehicle transient response test procedure (sinusoidal input)*, Draft proposal for an International Standard.

(6) P.S.FANCHER, R.L.NISONGER, C.B.WINKLER, AND K.GUO (1981) *Analytical comparisons of methods for assessing the transient directional response of automobiles*, 7th IAVSD Symposium.

(7) R.L.NISONGER AND P.S.FANCHER *Experimental examination of Transient directional response tests*, SAE 810808.

(8) M.K.VERMA (1981) Transient response test procedures for measuring vehicle directional control, *Vehicle System Dynamics*, **10**(6).

CHAPTER SIX

Control and Stability of Articulated Vehicles

6.1 INTRODUCTION

Articulated semi-trailer vehicles which are widely employed for the transport of bulk goods have characteristics that are very different from those of passenger cars. For example the steady state lateral acceleration rarely exceeds 0.4g because the unit will roll over due to the ratio of centre of gravity height to track width. There are a number of unstable motions which may be observed; the tractor can be described as under or oversteer, a snaking oscillation can be developed by the semi-trailer, the tractor may jack-knife either under power or when braking and trailer swing occurs when the trailer axle wheels are locked during braking.

In this chapter these vehicles will be treated as two bodies constrained together at the coupling or fifth wheel which provides a convenient origin for the equations of motion. A four degree of freedom model is developed which has forward speed, lateral velocity, yaw velocity and articulation between tractor and trailer but does not consider roll. This model will show how jack-knifing and trailer swing occur on low friction surfaces and is a useful tool for examining braking ratios, brake timing and other phenomena.

A linear model will be derived from the four degree of freedom system to give the responses to steering, wind gusts and road camber. As in the study of rigid vehicles this linear system is also used to examine the gains, frequencies and damping of the non linear models employed for complex computerised analyses.

Most semi-trailer vehicles have twin tyres and multi axle bogies at the rear tractor axle and also at the semi trailer axle. As a result of this there are 'locked in' torques because both axles cannot take up the ideal attitude angle and thus lateral forces are developed within the axle set even at very low speeds. A self steering trailer axle which attempts to relieve this situation is described.

Full trailer and multiple body vehicles are less widely employed than the semi-trailer; models of these units are described together with some of the instabilities encountered in their operation.

6.2 THE TYPICAL FIFTH WHEEL

A sketch of a fifth wheel is shown in Fig.6.1. The kinematic design of the fifth wheel provides for free articulation in the yaw mode while locating the tractor and trailer in longitudinal and lateral directions.

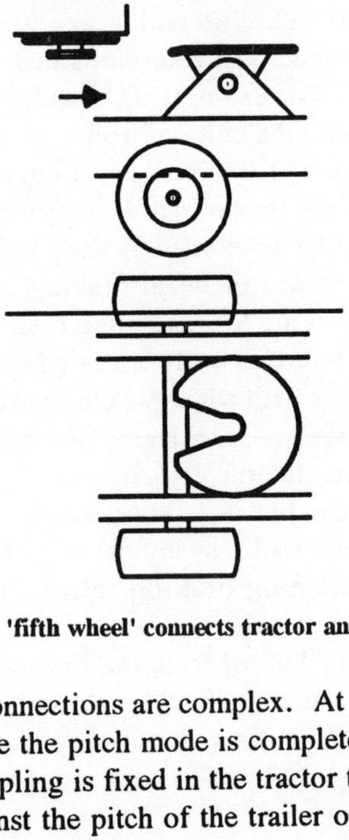

Fig.6.1 The 'fifth wheel' connects tractor and semi-trailer

The pitch and roll connections are complex. At zero articulation angle a roll restraint exists while the pitch mode is completely free. However since the pitch axis of the coupling is fixed in the tractor the roll movement of the tractor starts to act against the pitch of the trailer once an articulation angle is present.

When the articulation angle is ninety degrees roll of the tractor is effectively restrained by the trailer. Conversely the roll restraint of the trailer decreases with articulation angle until at ninety degrees the fifth wheel makes no contribution to roll stiffness of the trailer. The situation is further complicated by the fact that in many rigs wear of the pin and plates allows backlash.

6.3 DIMENSIONS

The typical symbols used to represent the dimensions of the articulated vehicle are shown in Fig.6.2. It is convenient to locate the origin for both tractor and semi trailer at the common point of the fifth wheel coupling and all measurements are made from this point. The symbols a and b define the front and rear axles with subscripts 1; e.g., a_1, b_1 for the tractor axles and b_2 for the semi-trailer axle. In the case of a multi-axle bogie then the measurement is taken to the centre of the axles and the spread is $2e$.

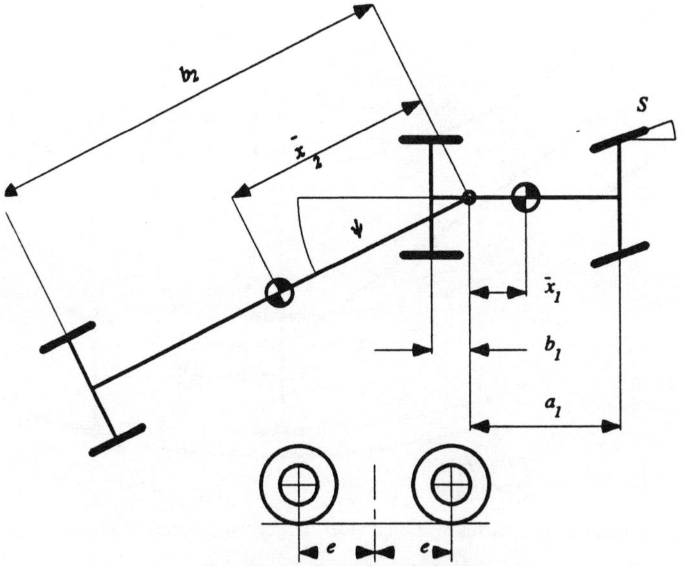

Fig.6.2 The dimensions and symbols used in articulated vehicle models

The centres of gravity are denoted by $\bar{x}$ with subscript 1 or 2 . The track of an axle is y_f , y_r , and y_t. The articulation angle between the units is positive in the direction shown.

6.4 LOW SPEED STEERING AND ARTICULATION

At low forward speeds the effects of lateral acceleration are low and the attitudes of the semi-trailer vehicle are governed by kinematic considerations.

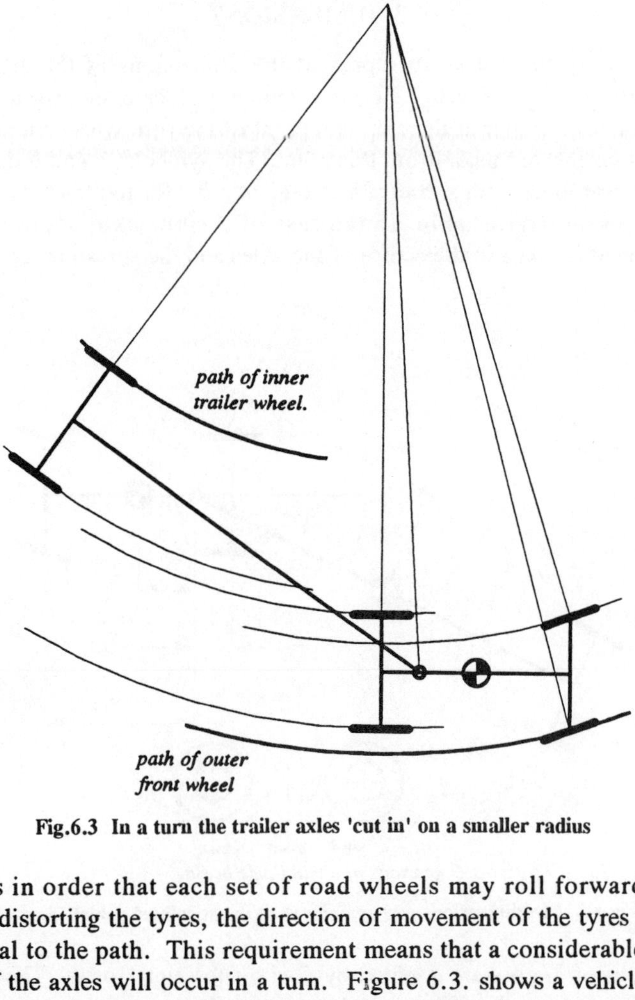

Fig.6.3 In a turn the trailer axles 'cut in' on a smaller radius

Thus in order that each set of road wheels may roll forward freely without distorting the tyres, the direction of movement of the tyres must be tangential to the path. This requirement means that a considerable lateral offset of the axles will occur in a turn. Figure 6.3. shows a vehicle at low speed on a curved path, the difference in tracking between tractor and trailer is clearly visible. Regulations in some countries require that the whole vehicle is capable of manoeuvering in a sharp turn between specified radii, or within a given width of traffic lane. Typical dimensions for the legal inner and outer radii are 6.3 and 12 metres respectively. This type of diagram is useful in demonstrating that capability and may also show how a steering axle can change cut in.

6.5 FORCE BALANCE IN A STEADY-STATE TURN

Tyre forces and moments may be calculated for the steady state turn by considering the lateral equilibrium and moment balance for the trailer and tractor as separate units. The connection at the fifth wheel is replaced by lateral forces, the balance of these forces provides the additional equation required to complete the computation.

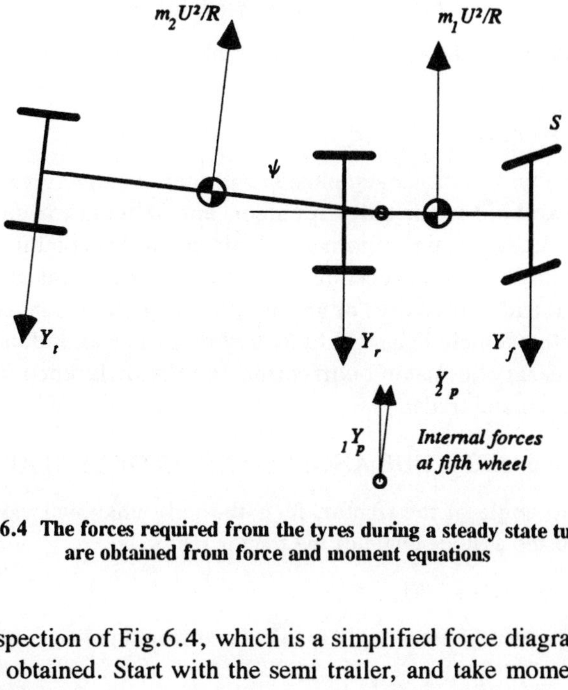

Fig.6.4 The forces required from the tyres during a steady state turn are obtained from force and moment equations

From inspection of Fig.6.4, which is a simplified force diagram, a set of equations is obtained. Start with the semi trailer, and take moments around the fifth wheel

$$Y_t = (m_2\, U^2/R)\bar{x}_2\, /b_2 \qquad (6.1)$$

The balance of lateral forces for the trailer yields the value of the internal force between tractor and trailer

$${}_2Y_p = m_2\, U^2/R - Y_t \qquad (6.2)$$

or

$${}_2Y_p = m_2\, U^2/R(1-\bar{x}_2\, /b_2)$$

Equilibrium of the fifth wheel gives the effective lateral force transmitted from trailer to tractor.

$$ {}_1Y_p = -{}_2Y_p \cos\psi \qquad (6.3) $$

The front axle of the tractor is a convenient position around which to take moments.

$$ Y_r = \{m_1\, U^2/R(a_1 - \bar{x}_1) - {}_1Y_p\, a_1\}/(a_1 + b_1) \qquad (6.4) $$

Finally, summing lateral forces on the tractor gives the lateral force at the front axle.

$$ Y_f = m_1\, U^2/R - Y_r - {}_1Y_p \qquad (6.6) $$

From the tyre characteristic plots the attitude angles at each axle may be calculated and hence the sideslip, steer, and articulation angles for the vehicle on a known radius of turn are obtained. At this point it is possible to make an estimate of the real centre of rotation and hence calculate new radii for the front and rear centres of gravity, the calculation converges rapidly and unless the vehicle is being studied while in a small radius turn it may not be necessary to make a correction for the difference in turn radii between tractor and trailer.

6.6 ATTITUDE ANGLES AND ARTICULATION

The sideslip angle of the tractor, ß, is the only unknown variable in the equation for the attitude angle of the tractor rear axle.

$$ \alpha_r = -(\beta - b_1/R) \qquad (6.6) $$

When ß is known the steer angle is calculated from the front attitude angle.

$$ \alpha_f = s - (\beta + a_1/R) \qquad (6.7) $$

The trailer articulation angle comes from the attitude angle of the semi trailer tyres.

$$ \alpha_t = \psi - (\beta - b_2/R) \qquad (6.8) $$

6.7 KINEMATICS OF THE FIFTH WHEEL

The fifth wheel is the point at which tractor and trailer are connected, hence the velocity of the fifth wheel must be the same whether it is described in terms of tractor or trailer variables. From Fig.6.5 it is seen that

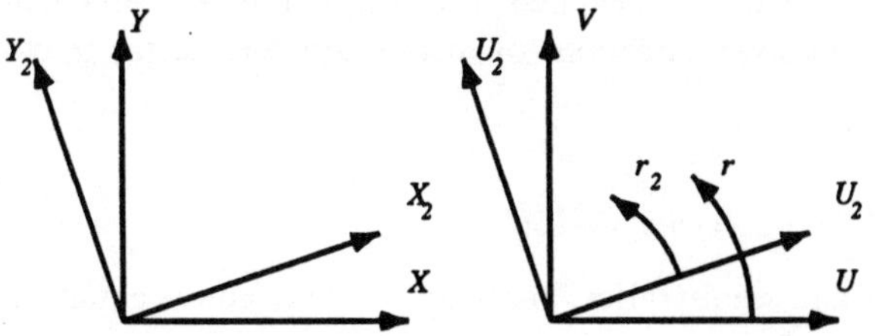

Fig.6.5 Equilibrium of forces & the relations between variables at the fifth wheel

$$U_2 = U\cos\psi + V\sin\psi$$
$$V_2 = -U\sin\psi + V\cos\psi \qquad (6.9)$$

where,

$$\psi = \int (r_2 - r)dt$$

Differentiation of equation (6.9) gives the accelerations of the trailer in terms of the tractor variables

$$\dot{U}_2 = \{\dot{U} + V(r_2 - r)\}\cos\psi + \{\dot{V} - U(r_2 - r)\}\sin\psi$$
$$\dot{V}_2 = -\{\dot{U} + V(r_2 - r)\}\sin\psi + \{\dot{V} - U(r_2 - r)\}\cos\psi \qquad (6.10)$$

6.8 EQUILIBRIUM OF FORCES AT THE FIFTH WHEEL

The rotation matrix also provides the relations between the internal forces at the fifth wheel. These forces act in the directions of the longitudinal and lateral axes of tractor and semi-trailer, thus

$${}_2X_p = X_p\cos\psi + Y_p\sin\psi$$
$${}_2Y_p = -X_p\sin\psi + Y_p\cos\psi \qquad (6.11)$$

6.9 A SIDESLIP YAW VELOCITY AND ARTICULATION ANGLE MODEL

A model of the semi-trailer vehicle with constant forward speed and sideslip, yaw velocity and articulation angle as the variables provides a means of examining many of the problems of articulated vehicle stability. Fore/aft equilibrium is assumed. From equation (2.7) in Chapter Two the lateral and rotational equations of motion are derived for tractor and semi-trailer.

$$\Sigma Y = m(\dot{V}+Ur) + m\bar{x}\dot{r}$$
$$\Sigma N = I_z\dot{r} + m\bar{x}(\dot{V}+Ur) \qquad \text{from (2.7)}$$

For tractor and semi-trailer, application of the above equations yield four equations

$$m_1(\dot{V}+Ur) + m_1\bar{x}_1\dot{r} = Y_f + Y_r + {}_1Y_p$$
$${}_1I_z\dot{r} + m_1\bar{x}_1(\dot{V}+Ur) = aY_f - bY_r$$
$$m_2(\dot{V}_2+Ur_2) - m_2\bar{x}_2\dot{r}_2 = Y_t + {}_2Y_p$$
$${}_2I_z\dot{r}_2 - m_2\bar{x}_2(\dot{V}_2+Ur_2) = -b_2Y_t$$

From equation (6.10)

$$\dot{V}_2+Ur_2 \approx \dot{V}+Ur$$

and

$${}_1Y_p \approx -{}_2Y_p$$

When the internal forces at the fifth wheel are eliminated the following equations remain.

$$m(\dot{V}+Ur) + m_1\bar{x}_1\dot{r} - m_2\bar{x}_2\dot{r}_2 = Y_f + Y_r + Y_t$$
$$m_1\bar{x}_1(\dot{V}+Ur) + {}_1I_z\dot{r} = a_1Y_f - b_1Y_r$$
$$-m_2\bar{x}_2(\dot{V}+Ur) + {}_2I_z\dot{r}_2 = -b_2Y_t$$

The linearised tyre forces are written in terms of these variables with the assumption of a common attitude angle for both the sets of tyres on a single axle.

$$\text{Trailer} \quad Y_f = C_f\{s - (V+a_1r)/U\}$$
$$Y_r = C_r\{-(V-b_1r)/U\} \qquad (6.12)$$

$$\text{Trailer} \quad Y_t = C_t\{-(V_2 - b_2r_2)/U_2\}$$

$$\text{or} \quad Y_t = C_t(-U\sin\psi + V\cos\psi - b_2r_2)/(U\cos\psi + V\sin\psi)$$

$$\text{thus} \quad Y_t = C_t(-(V - b_2r_2))/U \qquad (6.13)$$

When equations (6.12) and (6.13) are combined a set of linear equations which describe the basic handling characteristics of an articulated semi-trailer vehicle is obtained. These equations are manipulated in the same way as the linear equations of the car models to give transient and steady state responses.

$$\begin{bmatrix} m & m_1\bar{x}_1 & -m_2\bar{x}_2 \\ m_1\bar{x}_1 & {}_1I_z & 0 \\ -m_2\bar{x}_2 & 0 & {}_2I_z \end{bmatrix} \begin{bmatrix} \dot{v} \\ \dot{r} \\ \dot{r}_2 \end{bmatrix}$$

$$= \begin{bmatrix} -(C_f+C_r+C_t)/U & -mU-(a_1C_f-b_1C_r-b_2C_t)/U & b_2C_t/U & C_t & C_f \\ -(a_1C_f-b_1C_r)/U & -m_1\bar{x}_1U-(a_1^2C_f+b_1^2C_r)/U & 0 & 0 & a_1C_f \\ b_2C_t/U & m_2\bar{x}_2U & -b_2^2C_t/U & -b_2C_t & 0 \end{bmatrix} \begin{bmatrix} v \\ r \\ r_2 \\ \psi \\ s \end{bmatrix}$$

Matrix 6.1 The linear articulated vehicle - equation (6.14)

6.10 STEADY STATE STEERING RESPONSES

When the vehicle achieves a steady state condition after a steering input all derivatives of variables become zero, also $r_2 = r$. Thus the steady state responses to steering are obtained from matrix 6.2. which is a re-arrangement of the right hand side of matrix 6.3. The steady state responses are obtained from the following equations by Cramer's Rule, always provided that the Jacobean is not zero.

$$\begin{bmatrix} C_f+C_r+C_t & mU+(a_1C_f-b_1C_r-b_2C_t)/U & -C_t \\ a_1C_f-b_1C_r & m_1\bar{x}_1U+(a_1^2C_f+b_1^2C_r)/U & 0 \\ -b_2C_t & -m_2\bar{x}_2U+b_2^2C_t/U & b_2C_t \end{bmatrix} \begin{bmatrix} \beta \\ r \\ w \end{bmatrix} = \begin{bmatrix} C_f \\ a_1C_f \\ 0 \end{bmatrix} s$$

Matrix 6.2 The steady state steering response equations for the linear semi-trailer vehicle - equation (6.15). Note that $r_2 = r$

Sideslip response

$$\beta/s\big|_{ss} = \frac{\begin{vmatrix} C_f & mU+(a_1C_f-b_1C_r-b_2C_t)/U & -C_t \\ a_1C_f & m_1\bar{x}_1U+(a_1^2C_f+b_1^2C_r)/U & 0 \\ 0 & -m_2\bar{x}_2U+b_2^2C_t/U & b_2C_t \end{vmatrix}}{\begin{vmatrix} C_f+C_r+C_t & mU+(a_1C_f-b_1C_r-b_2C_t)/U & -C_t \\ a_1C_f-b_1C_r & m_1\bar{x}_1U+(a_1^2C_f+b_1^2C_r)/U & 0 \\ -b_2C_t & -m_2\bar{x}_2U+b_2^2C_t/U & b_2C_t \end{vmatrix}} \tag{6.16}$$

Yawing velocity response

$$r/s\big|_{ss} = \frac{\begin{vmatrix} C_f+C_r+C_t & C_f & -C_t \\ a_1C_f-b_1C_r & a_1C_f & 0 \\ -b_2C_t & 0 & b_2C_t \end{vmatrix}}{\begin{vmatrix} C_f+C_r+C_t & mU+(a_1C_f-b_1C_r-b_2C_t)/U & -C_t \\ a_1C_f+b_1C_r & m_1\bar{x}_1U+(a_1^2C_f+b_1^2C_r)/U & 0 \\ -b_2C_t & -m_2\bar{x}_2U+b_2^2C_t/U & b_2C_t \end{vmatrix}} \tag{6.17}$$

Articulation angle response to steering

$$\psi/s\big|_{ss} = \frac{\begin{vmatrix} C_f+C_r+C_t & mU+(a_1C_f-b_1C_r-b_2C_t)/U & C_f \\ a_1C_f+b_1C_r & m_1\bar{x}_1U+(a_1^2C_f+b_1^2C_r)/U & a_1C_f \\ -b_2C_t & -m_2\bar{x}_2U+b_2^2C_t/U & 0 \end{vmatrix}}{\begin{vmatrix} C_f+C_r+C_t & mU+(a_1C_f-b_1C_r-b_2C_t)/U & -C_t \\ a_1C_f+b_1C_r & m_1\bar{x}_1U+(a_1^2C_f+b_1^2C_r)/U & 0 \\ -b_2C_t & -m_2\bar{x}_2U+b_2^2C_t/U & b_2C_t \end{vmatrix}} \tag{6.18}$$

6.11 CRITICAL/CHARACTERISTIC SPEED OF THE ARTICULATED VEHICLE

The characteristic steady state behaviour of the system is described by the determinant of the left hand square matrix of matrix 6.3.

$$\begin{vmatrix} C_f+C_r+C_t & mU+(a_1C_f-b_1C_r-b_2C_t)/U & -C_t \\ a_1C_f+b_1C_r & m_1\bar{x}_1U + (a_1^2C_f+b_1^2C_r)/U & 0 \\ -b_2C_t & -m_2\bar{x}_2U+b_2^2C_t/U & b_2C_t \end{vmatrix}$$

Expansion of the characteristic determinant will show that the articulated semi trailer can be stable or divergently unstable depending upon the tyres and dimensions. The following manipulations ease the task of expanding the determinant.

$$\textit{column 1} = \textit{column 1} + \textit{column 3}$$

$$\textit{column 2} = \textit{column 2} + b_2 \cdot \textit{column 3}$$

Thus

$$\begin{vmatrix} C_f+C_r & mU+(a_1C_f-b_1C_r)/U & -C_t \\ a_1C_f-b_1C_r & m_1\bar{x}_1U + (a_1^2C_f+b_1^2C_r)/U & 0 \\ 0 & -m_2\bar{x}_2U & b_2C_t \end{vmatrix}$$

or

$$\begin{aligned} &- m_1 Ub_2 C_t \{C_f (a_1 - \dot{x}_1)-C_r (b_1+\dot{x}_1)\} \\ &\qquad - m_2 UC_t (b_2 - \dot{x}_2)(a_1 C_f - b_1 C_r)+b_2 L^2C_f C_r C_t /U \end{aligned}$$

The vehicle will be understeering and stable at all speeds provided that

$$\begin{aligned} &-m_1 U^2b_2 \{C_f (a_1 - \dot{x}_1) - C_r (b_1+\dot{x}_1)\} \\ &\qquad - m_2 U^2(b_2 - \dot{x}_2)(a_1 C_f - b_1 C_r)+b_2 L^2C_f C_r \geq 0 \end{aligned} \qquad (6.19)$$

This implies that the two terms containing the tractor and semi-trailer masses must together be positive for understeer.

$$-m_1 b_2 \{C_f (a_1 - \dot{x}_1) - C_r (b_1+\dot{x}_1)\} - m_2 (b_2 - \dot{x}_2)(a_1 C_f - b_1 C_r) \geq 0$$

Otherwise the semi trailer vehicle is stable at speeds below the critical speed and unstable at higher speeds. The critical speed for the oversteer vehicle is given when equation (6.19) is zero.

$$U^2 = \frac{b_2 L^2C_f C_r}{m_1 b_2 \{C_f (a_1 - \dot{x}_1)-C_r (b_1+\dot{x}_1)\} + m_2 (b_2 - \dot{x}_2)(a_1 C_f - b_1 C_r)} \qquad (6.20)$$

6.12 CONSTANT RADIUS RESPONSES

For the basic handling test of driving on a fixed radius at various speeds the steering response matrix is modified as shown in matrix 6.3.

$$\begin{bmatrix} C_f+C_r+C_t & -C_f & -C_t \\ a_1C_f-b_1C_r & -a_1C_f & 0 \\ b_2C_t & 0 & -b_2Ct \end{bmatrix} \begin{bmatrix} \beta \\ s \\ \psi \end{bmatrix} = \begin{bmatrix} -mU^2-(a_1C_f-b_1C_r-b_2C_t) \\ -m_1\bar{x}_1U^2-(a_1{}^2C_f+b_1{}^2C_r) \\ m_2\bar{x}_2U^2-b_2{}^2C_t \end{bmatrix} 1/R$$

Matrix 6.3 The steady state curvature responses for the linear semi-trailer vehicle

The sideslip angle response is given by the following equation.

$$\beta = \frac{\begin{vmatrix} -mU^2-(a_1C_f-b_1C_r-b_2C_t) & -C_f & -C_t \\ -m_1\bar{x}_1U^2-(a_1^2C_f+b_1^2C_r) & -a_1C_f & 0 \\ m_2\bar{x}_2U^2-b_2^2C_t & 0 & -b_2Ct \end{vmatrix} 1/R}{\begin{vmatrix} C_f+C_r+C_t & -C_f & -C_t \\ a_1C_f-b_1C_r & -a_1C_f & 0 \\ b_2C_t & 0 & -b_2Ct \end{vmatrix}} \tag{6.21}$$

Steer angle

$$s = \frac{\begin{vmatrix} C_f+C_r+C_t & -mU^2-(a_1C_f-b_1C_r-b_2C_t) & -C_t \\ a_1C_f-b_1C_r & -m_1\bar{x}_1U^2-(a_1^2C_f+b_1^2C_r) & 0 \\ b_2C_t & m_2\bar{x}_2U^2-b_2^2C_t & -b_2Ct \end{vmatrix} 1/R}{\begin{vmatrix} C_f+C_r+C_t & -C_f & -C_t \\ a_1C_f-b_1C_r & -a_1C_f & 0 \\ b_2C_t & 0 & -b_2Ct \end{vmatrix}} \tag{6.22}$$

Articulation angle

$$\psi = \frac{\begin{vmatrix} C_f+C_r+C_t & C_f & -mU^2-(a_1C_f-b_1C_r-b_2C_t) \\ a_1C_f-b_1C_r & a_1C_f & -m_1\bar{x}_1U^2-(a_1^2C_f+b_1^2C_r) \\ b_2C_t & 0 & m_2\bar{x}_2U^2-b_2^2C_t \end{vmatrix} 1/R}{\begin{vmatrix} C_f+C_r+C_t & -C_f & -C_t \\ a_1C_f-b_1C_r & -a_1C_f & 0 \\ b_2C_t & 0 & -b_2Ct \end{vmatrix}} \qquad (6.23)$$

The position of the fifth wheel and the location of the trailer load are significant factors in the design and operation of semi-trailer vehicles. As the fifth wheel is moved forward on the tractor the unit becomes more understeer and also shows less tendency to jack-knife under braking. The disposition of the load on the semi-trailer has a great influence on the control sensitivity of the rig. With the disposable load on the semi-trailer in the forward position the vehicle is tends to oversteering. As the load is shifted toward the trailer axle the handling changes to understeering. The reason for this change is clearly the increased demand for lateral force from the tractor rear tyres when the load is in the forward position.

6.13 DYNAMIC RESPONSES; ROOTS OF EQUATIONS OF MOTION

When the equations of motion are linearised they may be applied at any steady state condition to examine the local responses and stability of the system. One method of looking at the characteristics is to find the numerical roots of the equations using one of the root finding techniques available for digital computers. The characteristic equation is given below. An initial estimate of the roots is obtained from the terms in the leading diagonal, these values may be refined using Newton - Raphson or similar methods. This method has been employed for a hypothetical small semi-trailer vehicle with the following results.

$$\begin{vmatrix} [mD+(C_f+C_r+C_t)/U] & [(m_1\bar{x}_1-m_2\bar{x}_2)DU+mU+(a_1C_f-b_1C_r-b_2C_t)/U] & [(-m_2\bar{x}_2)D^2-(b_2C_t/U)D+C_t] \\ [m_1\bar{x}_1D+(a_1C_f-b_1C_r)/U] & [{}_1I_zD+m_1\bar{x}_1U+(a_1^2C_f+b_1^2C_r)/U] & 0 \\ [-m_2\bar{x}_2D-b_2C_t/U] & [{}_2I_zD-m_2\bar{x}_2U+b_2^2C_t/U] & [{}_2I_zD^2+(b_2^2C_t/U)D+b_2C_t] \end{vmatrix} = 0$$

The calculations are made for two forward speeds.

The numerical data used in determining the roots of the characteristic equation is shown in Table 6.1.

Table 6.1

	Tractor	*Semi-trailer*
m (kg)	2700	9000
I_z (kg/m^2)	4080	47 500
Wheelbase (m)	2.8	7
Centre of gravity (m) (from front)	1.2	?
C_f (N/rad)	104 600	
C_r	308 000	259 000

Figure 6.6 shows that the position of the fifth wheel and the location of the centre of gravity of the load both affect the handling responses. At the low speed of 20 mile/h with the fifth wheel well forward of the rear axle and the load well forward on the trailer all the roots are negative and the response times are short. As the load is shifted to the rear of the trailer it first appears that the trailer will develop a damped oscillatory response, shown by the appearance of complex roots. This oscillation is well damped with a time period less than 2 seconds. Further rearward displacement of the load gives rise to two pairs of complex roots, those relating to the tractor are less damped and have a longer time period, exceeding five seconds, than the trailer.

At the higher speed the roots tend to be complex with the forward position of the fifth wheel and any position of the trailer centre of gravity.

20 mph.

C.G of semi-trailer & load.	Tractor C of G. 1.2	1.8 m	2.4
1.5	-8.38 -1.6 # -13.6 -6.0		
3 m	-6 -1.8 # -9.3+/-3.2i	-1.3 # -9.3 -6.3+/-1.1i	-.56 # -11 -5+/-0.5i
4.5	-2.2+/-1.2i -11.2+/-3.9i		
6	-1.2+/-1.5i -13.3+/-3i		

40 mph.

C.G of semi-trailer & load.	Tractor C of G. 1.2	1.8 m	2.4
1.5	-2.8 # -4.0 -4+/-3.4i		
3 m	-2.3+/-2.1i -4.3+/-4.3i	-.01 # -6 -2.7+/-3.7i	1.3 !! -7.2 -2.4+/-3.6i
4.5	-1.1+/-2.2i -5.6+/-5.2i		
6	-.6+/-2i -6.8+/-5.8i		

Fig.6.6 The roots of the characteristic equation.
*** smallest negative root. ! instability**

As the fifth wheel is moved rearward in the tractor the system tends to instability and one example of numerical instability is seen.

6.14 CAR AND TRAILER COMBINATIONS

The car and trailer used for recreational purposes is an example of an articulated vehicle similar in concept to the models developed here. In particular the ball hitch or fifth wheel lies behind the rear axle of the car and the centre of gravity of the trailer is near the trailer axle. From the table of roots of the characteristic equation it is reasonable to make the assessment that the car and trailer may be unstable, most probably in an oscillatory mode. A study using the technique of root computation provides some conclusions regarding the passenger car/trailer home combination. These are described in terms of the static and the oscillatory stability. Static stability, given by real roots, should be positive and of sufficiently short response time that the driver is not required to learn new steering tasks when towing. Oscillations which are lightly damped can be difficult for the driver.

The trailer weight should be small, increasing the mass has adverse effects on both the static and oscillatory responses. Increasing the length of the draw-bar has a moderately benificial effect on both static and oscillatory stability. Moving the centre of gravity forward has a severe adverse effect on the static stability while improving the oscillatory behaviour.

A long wheelbase car with a short overhang to the fifth wheel is the best towing vehicle. The statement in the previous paragraph describing the forward movement of the trailer load as reducing the static stability is probably due to the increase in load on the rear axle and corresponding reduction in load on the front axle of the towing vehicle which will increase the oversteering tendency.

6.15 A NON-LINEAR SEMI-TRAILER MODEL

When forward speed is included as a variable, a large articulation angle is permitted and non linear tyre characteristics are introduced a model capable examining many of the problems of operating these vehicles in adverse conditions is available. The original analysis was developed in the early 1960s and used to study jack-knifing, trailer swing, and other non-linear events. The various stages in the development of a simulation are now presented. It starts with a non linear system in which the tractor and trailer are assigned global variables, the *state variables*, which are then related by the kinematic restraints of the fifth wheel. This system of equations may be solved with the aid of a digital computer but does not provide an engineering insight into the chassis design process. The second step is the elimination of the redundant variables and the internal forces at the fifth wheel thus reducing the model to a four degree of freedom system with forward speed, lateral velocity, yaw velocity and articulation angle as the variables.

The origin for the axis systems for both tractor and semi-trailer is located at the centre of the fifth wheel. The symbols U, V, r given in Fig.6.4 describe the tractor, while U_2, V_2, r_2 are the forward, lateral, and yaw velocities of the trailer.

6.16 EQUATIONS OF MOTION IN STATE VARIABLE NOTATION

The equations of motion are written with the origin for both tractor and trailer located at the common point of the fifth wheel pivot. The centre of gravity of each unit is assumed to lie on the central axis of that vehicle. The wheels are numbered 1 and 2, 3 and 4, and 5 and 6. Left side wheels are odd numbered.

Tractor:

$$m_1(\dot{U}-Vr) - m_1\bar{x}_1 r^2 = X_1 + X_2 + X_3 + X_4 + X_p$$

$$m_1(\dot{V}+Ur) + m_1\bar{x}_1\dot{r} = Y_f + Y_r + Y_p$$

$$m_1\bar{x}_1(\dot{V}+Ur) + {}_1I_z\dot{r}$$

$$= a_1 Y_f - b_1 Y_r + N_f + N_r + (X_1 - X_2)y_f + (X_3 - X_4)y_r$$

Trailer: (6.23)

$$m_2(\dot{U}_2 - V_2 r_2) + m_2\bar{x}_2 r_2^2 = X_t + {}_2X_p$$

$$m_2(\dot{V}_2 + U_2 r_2) - m_2\bar{x}_2 r_2 = Y_t + {}_2Y_p$$

$$-m_2\bar{x}(\dot{V}_2 + U_2 r_2) + {}_2I_z\dot{r}_2 = -Y_t b_2 + N_t + (X_6 - X_5)y_t$$

The tyre attitude angles are written for the individual wheels in the states variable notation and the values for tyre lateral force are computed.

Tractor front axle:

$$\alpha_1 = s - (V + a_1 r)/(U + y_f r)$$

$$\alpha_2 = s - (V + a_1 r)/(U - y_f r)$$

Tractor rear axle:

$$\alpha_3 = -(V - b_1 r)/(U + y_r r)$$

$$\alpha_4 = -(V - b_1 r)/(U - y_r r)$$

Trailer axle:

$$\alpha_5 = -(V_2 - b_2 r_2)/(U_2 + y_t r_2)$$

$$\alpha_6 = -(V_2 - b_2 r_2)/(U_2 - y_t r_2)$$

The lateral tyre forces Y_n are computed as cubic functions of the attitude angles α_n. At this stage in the analysis it is possible to write these equations in matrix form and perform all the numerical manipulation on a digital

$$\begin{bmatrix} m_1 & 0 & 0 & 0 & 0 & 0 & -1 & 0 & 0 & 0 \\ 0 & m_1 & m_1\bar{x}_1 & 0 & 0 & 0 & 0 & -1 & 0 & 0 \\ 0 & m_1\bar{x}_1 & {}_1I_z & 0 & 0 & 0 & 0 & 0 & 0 & 0 \\ 0 & 0 & 0 & m_2 & 0 & 0 & 0 & 0 & -1 & 0 \\ 0 & 0 & 0 & 0 & m_2 & -m_2\bar{x}_2 & 0 & 0 & 0 & -1 \\ 0 & 0 & 0 & 0 & -m_2\bar{x}_2 & {}_2I_z & 0 & 0 & 0 & 0 \\ 0 & 0 & 0 & 0 & 0 & 0 & \cos\psi & \sin\psi & 1 & 0 \\ 0 & 0 & 0 & 0 & 0 & 0 & -\sin\psi & \cos\psi & 0 & 1 \\ -\cos\psi & \sin\psi & 0 & 1 & 0 & 0 & 0 & 0 & 0 & 0 \\ \sin\psi & \cos\psi & 0 & 0 & 1 & 0 & 0 & 0 & 0 & 0 \end{bmatrix} \begin{bmatrix} \dot{U} \\ \dot{V} \\ \dot{r} \\ \dot{U}_2 \\ \dot{V}_2 \\ \dot{r}_2 \\ X_p \\ Y_p \\ {}_2X_p \\ {}_2Y_p \end{bmatrix}$$

$$= \begin{bmatrix} m_1(Vr+\bar{x}_1 r^2)+X_1+X_2+X_3+X_4 \\ -m_1 Ur+Y_f+Y_r \\ a_1 Y_f - b_1 Y_r + N_f + N_r \\ m_2(V_2 r_2 - \bar{x}_2 r_2^2)+X_5+X_6 \\ -m_2 U_2 r_2 + Y_t \\ m_2\bar{x}_2 U_2 r_2 - b_2 Y_t + Nt \\ 0 \\ 0 \\ V_2(r_2-r)\cos\psi + U_2(r_2-r)\sin\psi \\ -V_2(r_2-r)\sin\psi + U_2(r_2-r)\cos\psi \end{bmatrix}$$

Matrix 6.4 States variable equations of motion for the semi-trailer vehicle with large articulation angle

computer. This approach is shown in matrix 6.4, where each of the original variables and the internal forces at the fifth wheel appear. The advantages of that procedure are that there is a very considerable reduction in the algebra and the variables appear in explicit form. For example, X_p and Y_p are directly available. The disadvantages are:

(a) the lack of 'feeling' for the solution that occurs when a large problem is approached without prior knowledge of the gains, frequencies and damping of the system;
(b) the questionable accuracy of continuous looping through large matrix operations followed by integration.

6.17 THE REDUCED EQUATIONS OF MOTION

By substituting the relations previously obtained for U_2, V_2, and their derivatives the system may be reduced to the following equations.

Tractor:

$$m_1(\dot{U}-Vr) - m_1\bar{x}_1 r^2 = X_1 + X_2 + X_3 + X_4 + X_p \qquad (a)$$

$$m_1(\dot{V}+Ur) + m_1\bar{x}_1\dot{r} = Y_f + Y_r + Y_p \qquad (b)$$

$$m_1\bar{x}_1(\dot{V}+Ur) + I_{1z}\dot{r}$$
$$= a_1Y_f - b_1Y_r + N_f + N_r + (X_1 - X_2)y_f + (X_3 - X_4)y_r \qquad (c)$$

Trailer: (6.24)

$$m_2\{(\dot{U}-Vr)\cos\psi + (\dot{V}+Ur)\sin\psi\} + m_2\bar{x}_2 r_2^2$$
$$= X_5 + X_6 - X_p\cos\psi - Y_p\sin\psi \qquad (d)$$

$$m_2\{(\dot{V}+Ur)\cos\psi - (\dot{U}-Vr)\sin\psi\} - m_2\bar{x}_2\dot{r}_2$$
$$= Y_t + X_p\sin\psi - Y_p\cos\psi \qquad (e)$$

$$-m_2\bar{x}_2\{(\dot{V}+Ur)\cos\psi - (\dot{U}-Vr)\sin\psi\} + I_{2z}\dot{r}_2$$
$$= -b_2Y_t + N_t + (X_5 - X_6)y_t \qquad (f)$$

Elimination of the forces at the fifth wheel gives four equations in terms of the variables U, V, R, and ψ. This is carried out by the as follows.

(1) Multiply equation (6.24d) by $\sin\psi$ and equation (6.24e) by $\cos\psi$, then add the resulting equations.

(2) Multiply equation (6.24e) by $\sin\psi$ and equation (6.24d) by $\cos\psi$, then subtract the equations.

(3) Add equations (6.24a) and (6.24h) to eliminate X_p, and equations 6.24(b) and (6.24g) to remove Y_p.

The equations of motion are now available in terms of the variables U, V, r, and ψ.

$$m(\dot{U}-Vr) + m_1\dot{x}_1\dot{r} + m_2\dot{x}_2(\dot{r}_2\sin\psi - r_2^2\cos\psi)$$
$$= X_1+X_2+X_3+X_4 + X_t\cos\psi - Y_t\sin\psi$$
$$m(\dot{V}+Ur) + m_1\dot{x}_1\dot{r} - m_2\dot{x}_2(\dot{r}_2\cos\psi + r_2^2\sin\psi)$$
$$= Y_f + Y_r + X_t\sin\psi + Y_t\cos\psi \qquad (6.25)$$
$$m_1\dot{x}_1(\dot{U}+Vr) + {}_1I_z\dot{r}$$
$$= a_1 Y_f - b_1 Y_r + N_f + N_r + (X_1 - X_2)y_f + (X_3 - X_4)y_r$$
$$-m_2\dot{x}_2\{(\dot{V}+Ur)\cos\psi - (\dot{U}-Vr)\sin\psi\} + {}_2I_z\dot{r}_2$$
$$= -b_2 Y_t + N_t + (X_5 - X_6)y_t$$

The lateral forces generated by the tyres are now written in terms of the same variables.

Tractor front axle:

$$\alpha_1 = s - (V + a_1 r)/(U + y_f r)$$
$$A\alpha\alpha_2 = s - (V + a_1 r)/(U - y_f r)$$

Tractor rear axle:

$$\alpha_3 = -(V - b_1 r)/(U + y_r r)$$
$$A\alpha\alpha_4 = -(V - b_1 r)/(U - y_r r)$$

Trailer axle:

$$\alpha_5 = -(-U\sin\psi + V\cos\psi - b_2 r_2)/(U\cos\psi + V\sin\psi + y_t r_2)$$
$$\alpha_6 = -(-U\sin\psi + V\cos\psi - b_2 r_2)/(U\cos\psi + V\sin\psi - y_t r_2)$$

With the equations in matrix format the inertial coupling between tractor and semi-trailer is clearly defined. The yawing moments arising from unbalanced longitudinal forces have been omitted from the matrix.

$$\begin{bmatrix} m & 0 & 0 & m_2\bar{x}_2 sin\psi \\ 0 & m & m_1\bar{x}_1 & -m_2\bar{x}_2 cos\psi \\ 0 & m_1\bar{x}_1 & {}_1I_z & 0 \\ -m_2\bar{x}_2 sin\psi & m_2\bar{x}_2 cos\psi & 0 & {}_2I_z \end{bmatrix} \begin{bmatrix} \dot{U} \\ \dot{V} \\ \dot{r} \\ \dot{r}_2 \end{bmatrix}$$

$$= \begin{bmatrix} mVr+m_1\bar{x}_1 r^2 - m_2\bar{x}_2 r_2^2 cos\psi + X_f + X_r + X_t cos\psi - Y_t sin\psi \\ -mUr - m_2\bar{x}_2 r_2^2 sin - X_t sin\psi + Y_f + Y_r + Y_t cos\psi \\ -m_1\bar{x}_1 Ur + a_1 Y_f - b_1 Y_r + N_f + N_r \\ m_2\bar{x}_2 r(Ucos\psi + Vsin\psi) - b_2 Y_t + N_t \end{bmatrix}$$

Matrix 6.5 The four-degree-of-freedom semi-trailer vehicle - equation (6.24)

Note that both the matrices contain terms which change with time. In the inertial matrix these time variant terms are those containing trigonometric functions. The left hand matrix terms include products of variables and also the tyre forces and moments which change as the vehicle responds to steering and braking. A digital simulation of this system requires that these terms are re-calculated at every time step, the matrix algebra must also be repeated.

6.18 APPLICATIONS FOR THE NON-LINEAR MODEL

There are no restrictions on the articulation angle or tyre attitude angles made in this analysis and thus the model may be used to study large angle movements of the rig which is particularly well suited for modeling the responses to braking and steering, delays in the response of trailer brakes and anti-lock brake systems.

The first use of the model was the simulation of braking in a turn. A typical vehicle is set up in a steady state turn when the brakes are applied to lock one of the wheel sets, the braking on the other axles is adjusted so that these wheels do not lock. When the tractor front axle wheels lock then steering control is lost and the vehicle moves in a straight path. While this condition is not dramatic the vehicle will soon leave the traffic lane and in the case of a cambered road will run into the berm. Violent jack-knifing occurs when the wheels on the tractor rear axles are locked, the vehicle rotates around the fifth wheel under the influence of the de-stabilising moment fron the front tyres. There is only a small lateral displacement of the fifth wheel because, particularly with a laden trailer, of the great mass of the trailer. With a typical braking set up jack-knifing is more likely to happen with an unladen trailer because in that condition the tractor rear axle takes too great a share of the braking.

When the trailer wheels lock the trailer swings outward on the curve. In many cases the operator is unaware of this since there is no indication given by changes in force or moment from the fifth wheel coupling.

6.19 DUAL TYRES

Where dual tyres are mounted on an axle there is a longitudinal force developed at each contact patch because the tyres are forced to rotate at a common angular velocity while travelling on paths of different curvature. Figure 6.7 illustrates the effect.

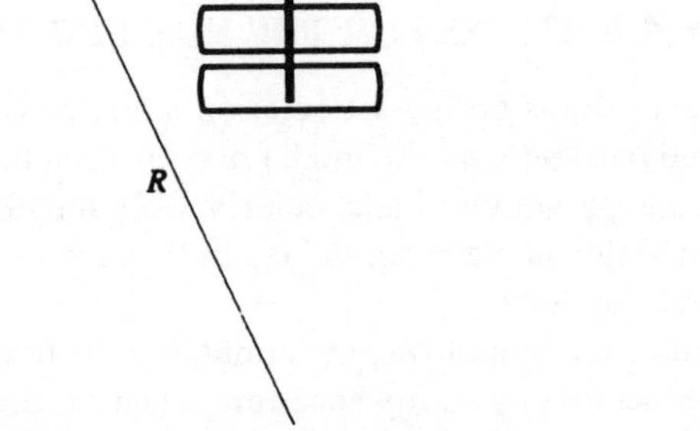

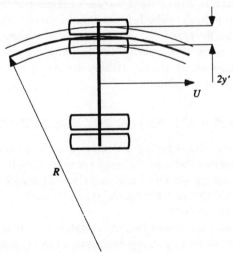

Fig.6.7 Dual tired wheels generate internal forces due to the incompatibility of kinematic rolling and slip angle

Let $2y'$ be the separation between the centres of the tyre contact patches, C_x is the slope of the longitudinal slip curve for a single tyre, that is the characteristic plot of longitudinal force versus slip of the tyre on the road where longitudinal slip is the ratio of actual speed at the contact surface to the free rolling speed.

Mean velocity of the wheel set

$$u = rR$$

The velocity on the larger radius is

$$u_o = r(R+y')$$

On the smaller radius

$$u_i = r(R-y')$$

Thus

$$X_o = -C_x(R+y')/R$$

and

$$X_i = C_x(R-y')/R$$

The moment from a pair of dual tyres is obtained by taking moments about the centre of the wheel set

$$N = -2C_x y'^2/R$$

In terms of speed and yaw velocity

$$N = -2C_x y'^2 r/U \qquad (6.27)$$

For the trailer the yaw velocity is the total yaw velocity and includes a component from the rate of change of articulation angle.

6.20 MULTI-AXLE SYSTEMS

Various types of multiple axle bogies are used on transport vehicles to increase the load capacity while limiting the force applied by any one axle to the road. In many cases the design of these devices results in large locked in forces due to the tyres working against each other in turning situations.

The simplest unit is the twin axle bogie without steering as shown in Fig.6.2. In a typical handling situation this will have an attitude angle at the centre of the axle spread which is a function of lateral velocity, forward speed, yaw velocity and, in the case of the trailer, the articulation angle and the angular velocity of articulation. It is these latter variables, the yaw and articulation angular velocities which cause the differences in attitude angles on a multi axle bogie. With an axle spread of $2e$ then the attitude angles at the front and rear axles of the bogie will differ from that at the centre by er/U, thus the effect on the tyre forces may be estimated.

For a typical rear axle displaced a distance 'b' from the origin the general expression for slip angle is

$$\alpha = -(V-br)/U$$

The individual axles will experience slip angles

$$\alpha = -\{V-(b+/-e)r\}/U$$

In the linear range of tyre characteristics the lateral forces generated by single tyres on each axle are

$$Y_1 = C\{-(V - (b-e)r)/U\}$$

$$Y_2 = C\{-(V - (b+e)r\}/U)$$

From the separation of the axles the moment is

$$N = e\,Y_1 - e\,Y_2$$

$$N = -\,2Ce^2r/U \qquad (6.28)$$

Tandem axles are frequently employed with dual tyres and then the moment locked into the suspension is the sum of these effects. The total moment is

$$N = -2(Ce^2 + C_x\,y'^2)r/U \qquad (6.29)$$

The analysis can be extended to multiple axle bogies used in some areas for special purposes vehicles.

6.21 THE SELF-STEERING AXLE

This device is an axle which permits the road wheels to steer under the action of a lateral tyre force. The steering is restrained by springs and dampers and is usually preloaded. Self steering appears to be used mainly on tandem and triple bogies where it is the rear axle. In some circumstances the axle may be prone to 'shimmy', a self-excited limit cycle vibration associated with steering, however this may be controlled by attention to the friction, preloading the system, the stiffness of the system and the supporting structure, reduction in the inertia of the system and sufficient damping. Figure 6.8. shows the basic design and the moments around the steering king pins in terms of the steer angle of the tyres.

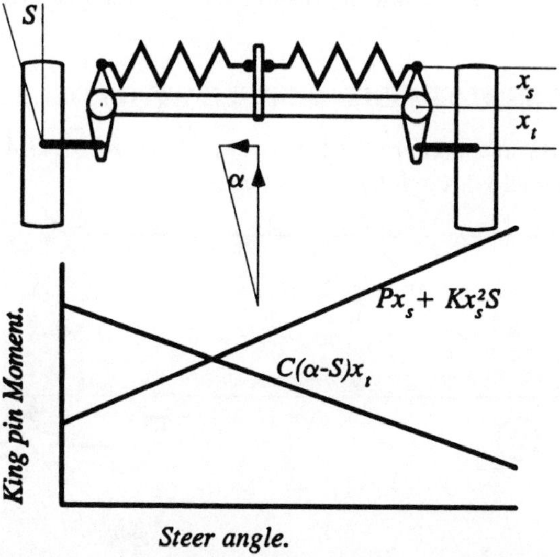

Fig.6.8 The self steering axle is steered in response to the moment around the steering axis from the lateral force and aligning moment of the tyre

For the present analysis the inertia of the system is ignored. Moment arms for the tyre and spring are designated by x_t and x_s, respectively. The spring stiffness is K , while the preload in the spring is P. The steer angle of the wheels is s, and the steering velocity is s,   is the slip angle of the vehicle at the position of the centre of contact of the tyres on the steering axle. The lateral force generated by the tyres is dependent upon the local slip angle, the steer angle and any lateral velocity generated by the longitudinal offset of the steering axis from the line of action of the lateral force. This line of action may be taken as not at the centre of contact if it is desired to account for aligning moment, otherwise the centre of contact is assumed.

$$Y = C(s - A - x_t s/U) \tag{6.30}$$

Consider the equilibrium of the steering arm around the pivot

$$Yx_t + (Kx_s^2 + Px_s)s = 0 \tag{6.31}$$

If the absolute value of Y is less than that required to overcome the preload then steering does not occur. When Y exceeds the preload then the system moves exponentially to a position of equilibrium. The similarity between this expression and the response of the laterally flexible tyre to steering is notable.

6.22 THE SEMI-TRAILER VEHICLE WITH ROLL FREEDOM

Roll moments are transmitted between the tractor and trailer of a commercial vehicle through the fifth wheel.

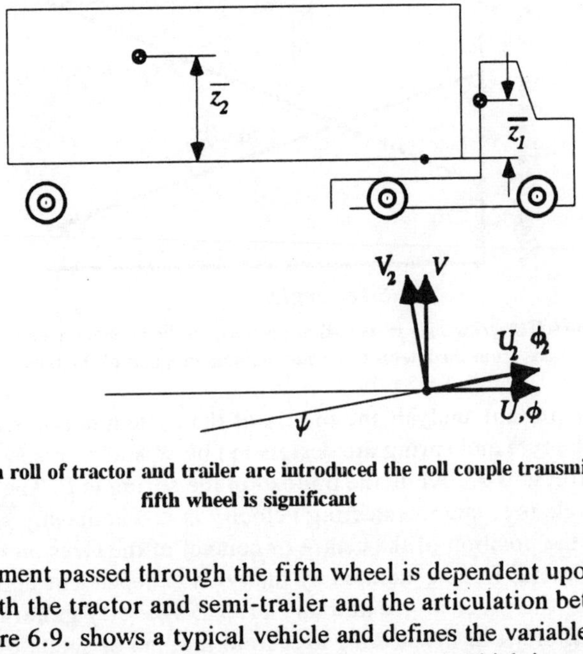

Fig.6.9 When roll of tractor and trailer are introduced the roll couple transmitted by the fifth wheel is significant

The moment passed through the fifth wheel is dependent upon the roll angle of both the tractor and semi-trailer and the articulation between the units. Figure 6.9. shows a typical vehicle and defines the variables. When the units are in line, zero articulation angle, a moment which is proportional to the difference in the roll angles of tractor and trailer acts in a spring like manner. At 90 degrees of articulation the front of the trailer is not restrained by the fifth wheel because the trailer axis now coincides with the pitch axis of fifth wheel. On the other hand the tractor is unable to roll

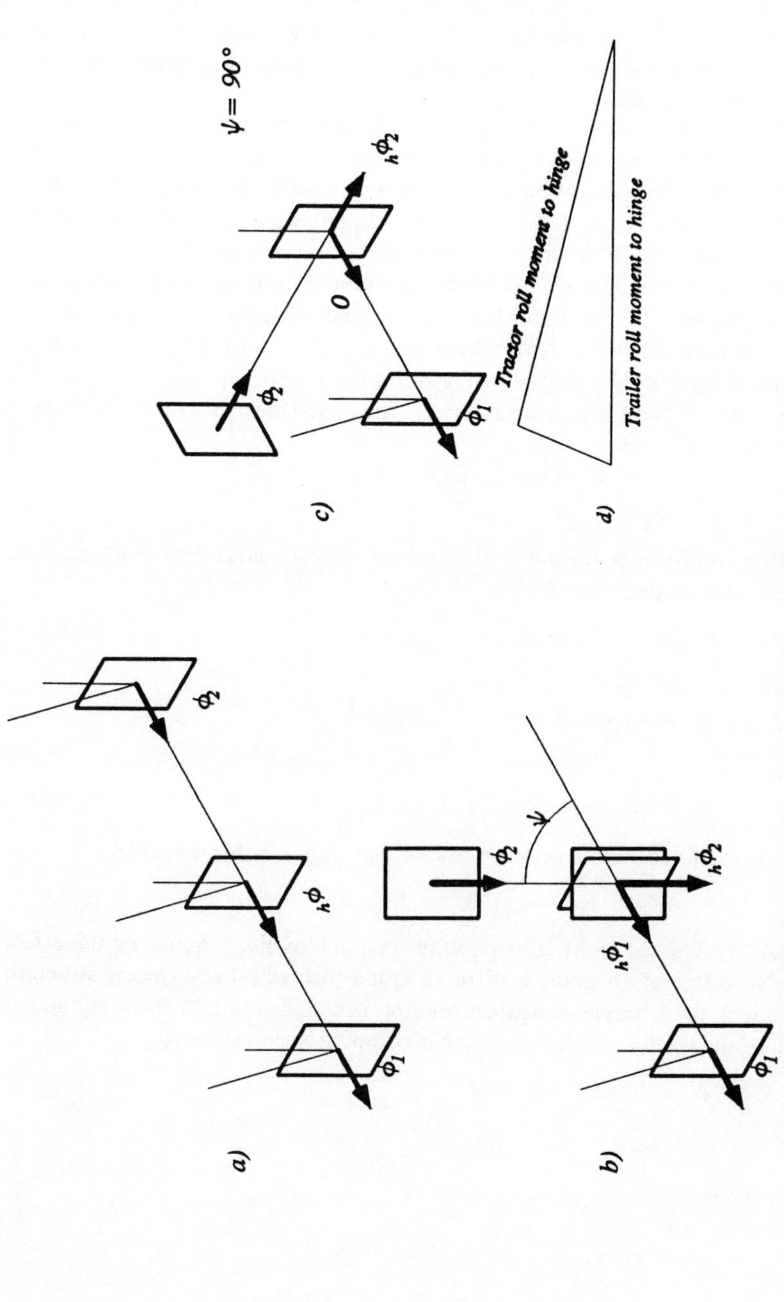

Fig.6.10 The roll moment transmitted by the fifth wheel of the semi-trailer vehicle is a function of articulation angle

because this motion is completely blocked by the trailer. Thus the effective stiffness of the fifth wheel appears to decrease as the articulation angle increases when viewed from the trailer, and to rise to an infinite value when seen from the tractor.

Some experimental fifth wheels have been tested in which the pitch axis is fixed relative to the trailer thus ensuring a rising roll stiffness for the trailer. After the publication of one report this work does not appear to have been continued. Figure 6.10. shows the roll angles of the tractor, trailer and at the plane of the fifth wheel for zero (a), an intermediate angle (b), and 90 degrees of articulation (c). A vector diagram (d) illustrates the partitioning of the moments on the trailer between roll and pitch for any articulation. Let the roll rotation of the fifth wheel be ${}_h\phi_1$ relative to the tractor and ${}_h\phi_2$ from the trailer. The stiffnesses relating the tractor and trailer to the fifth wheel are K_1 and K_2, respectively, these stiffnesses are the torsional stiffness of the structures.

Then, when the articulation angle is zero

$${}_h\phi_1 = {}_h\phi_2 = {}_h\phi$$

In this condition a torque is transmitted and the equilibrium at the fifth wheel may be determined

$$K_1 ({}_h\phi_1 - \phi_1) = K_2 (\phi_2 - {}_h\phi_2) \qquad (6.32)$$

Since

$${}_h\phi_1 = {}_h\phi_2 = {}_h\phi$$

then

$${}_h\phi_1 = (K_1 \phi_1 + K_2 \phi_2)/(K_1 + K_2) \qquad (6.33)$$

The transmitted torque between tractor and trailer is then obtained

$$L = K_1 K_2 (\phi_2 - \phi_1)/(K_1 + K_2) \qquad (6.34)$$

At 90 degrees of articulation with the pitch motion suppressed the effect on the tractor of assigning a stiffness to the fifth wheel and chassis structure is to lock the fifth wheel against rotation in the direction of the longitudinal axis of the tractor, while the trailer is completely unrestrained.

$${}_h\phi_1 = 0 \quad \text{and} \quad {}_h\phi_2 = \phi_2$$

$$L_f = K_1 \phi_1$$

$$L_r = 0$$

For small roll angles, since the fifth wheel is not restrained in pitch

$$ {}_h\phi_2 = {}_h\phi_1 \cos\psi + \phi_2 \sin\psi $$

From the vector diagram of Fig.6.10 the general case of equation (6.34) becomes

$$ K_1 ({}_h\phi_1 - \phi_1)\cos\psi = K_2 (\phi_2 - {}_h\phi_2) \qquad (6.35) $$

Thus the relative roll angles of the fifth wheel are obtained

$$ \begin{aligned} {}_h\phi_1 &= \{K_1 \phi_1 \cos\psi + K_2 \phi_2 (1-\sin\psi)\}/\{(K_1+K_2)\cos\psi\} \\ {}_h\phi_2 &= \{K_1 (\phi_1 \cos\psi + \phi_2 \sin\psi) + K_2 \phi_2\}/(K_1+K_2) \end{aligned} \qquad (6.36) $$

The transmitted moments are then

$$ \begin{aligned} L_1 &= K_1 ({}_h\phi_1 - \phi_1) \\ L_2 &= K_2 (\phi_2 - {}_h\phi_2) \end{aligned} \qquad (6.37) $$

For most cases of interest the articulation angle is small and the moment given by equation (6.34) is satisfactory.

It will be assumed that the relative roll angle is small for typical steering responses and the original equations relating the velocities of tractor and trailer (equations (6.4.)) are then applicable. The basic reason for considering the joint connecting the parts of the vehicle as a spring is that this represents the typical design of a commercial vehicle in which the frames are open sections joined by cross members so that the torsional stiffness of the resulting structure is low and the transfer of moments is accompanied by large angular movements. The equations of motion for the bodies are then the general expressions given below.

$$ \begin{aligned} X &= m(\dot{U}-Vr) - m\bar{x}r^2 + m\bar{z}pr \\ Y &= m(\dot{V}+Ur) + m\bar{x}\dot{r} - m\bar{z}\dot{p} \\ L &= I_x \dot{p} - m\bar{x}\bar{z}\dot{r} \\ N &= I_z \dot{r} - m\bar{x}\bar{z}\dot{p} \end{aligned} \qquad (6.38) $$

Specific values for the variables and parameters are substituted, the original axes for measurement of the inertial parameters are assumed to be principal axes, and the centres of gravity lie in the mid planes of the units.

A linearised vehicle running at steady forward speed is obtained with the usual small angle assumptions.

$$
\begin{aligned}
& m(\dot{V}+Ur) + m\bar{x}_1 \dot{r} - m\bar{z}_1 \dot{p} - m\bar{x}_2 \dot{r}_2 - m\bar{z}_2 \dot{p}_2 = Y_f + Y_r + Y_t \\
& mx_1 (\dot{V}+Ur) + {}_1I_z \dot{r} - m\bar{x}\bar{z}_1 \dot{p} = aY_f - bY_r \\
& -mz_1 (\dot{V}+Ur) - m\bar{x}\bar{z}_1 \dot{r} + {}_1I_x \dot{p} \\
& \qquad = -(C_{\phi f}+C_{\phi r})p - (K_{\phi f}+K_{\phi r})\phi \\
& \qquad\quad + gY_f + gY_r + K_1 ({}_h\phi_1 - \phi_1) \\
& -mx_2 (\dot{V}+Ur) + {}_2I_z \dot{r}_2 - m\bar{x}\bar{z}_2 \dot{p}_2 = -b_2 Y_t \\
& -mz_2 (\dot{V}+Ur) - m\bar{x}\bar{z}_2 \dot{r}_2 + {}_2I_x \dot{p}_2 \\
& \qquad = -C_{\phi t} - K_{\phi t}\phi + gY_t + K_2 (\phi_2 - {}_h\phi_2)
\end{aligned}
\tag{6.39}
$$

Note that the symbol $g = dy'/d\phi$, is the tyre lateral scrub rate with roll; since most trucks use a beam axle it is reasonable to assume the roll centre is at axle height. The lateral tyre forces also contain the tyre lateral velocity as part of the slip angles. Roll steer factors, $e = ds/d\phi$, are also included.

$$
\begin{aligned}
& Y_f = C_f \{s + e_f \phi - (V + a_1 r + g_f p)/U\} \\
& Y_r = C_r \{e_r \phi - (V - b_1 r + g_r p)/U\} \\
& Y_t = C_t \{\psi + e_t \phi_2 - (V - b_2 r_2 + g_t p_2)/U\}
\end{aligned}
\tag{6.40}
$$

Table 6.2 gives some measured values for the roll stiffnesses of various components of the tractor and semi-trailer. In all cases the tractor frame stiffness in torsion is extremely low and it is reasonable to assume that there is not a great possibility of changing the handling characteristics by the method familiar to passenger car engineers; that is by increasing the roll stiffness of a suspension with the objective of changing the proportioning of the normal force between the wheels. One other result of the frame flexibility is that there is a phase and amplitude difference between front and rear roll angles and lateral accelerations dependent upon steering frequency.

The torsional stiffness of the semi-trailer unit is mainly concentrated in the body which is mounted on the frame and can vary from extremely stiff to negligible. Fifth wheel stiffness shows considerable variation which may be due to the mounting or the design of the device. The effective roll stiffness is a combination of tyre and suspension effects, methods of

Table 6.2 Component contributions to the roll stiffness for large semi-trailer vehicles (x10^6 Nm/rad)

Vehicle	*No.1*	*No.2*	*No.3*
Tractor			
Rear tyres	4.13	8.91	4.48
Rear suspension	0.73	1.69	1.07
Front tyres	0.3	0.42	0.51
Front suspension	0.17	2.23	2.24
Frame torsion	0.27	0.27	0.27
Fifth wheel	14.1	16.3	6.65
Semi-trailer			
Tyres	6.76	6.94	16.2
Suspension	2.7	2.78	2.56
Frame and body	1.35	100	100

calculating the stiffness as a set of springs in series are demonstrated in Chapter Three and used here to develop the data given in Table 6.3.

Table 6.3 Overall roll stiffnesses for the tyres and suspensions (x10^6 Nm/rad)

Tractor	Front axle	0.25	0.36	0.42
	Rear axle	0.62	1.42	0.86
Semi-trailer	Axle	1.93	1.98	2.2

6.24 TRACTOR-TRAILER VEHICLES

The tractor-trailer rig usually consists of a tractor with the configuration of a rigid truck and the capability of carrying a commercial payload connected by a drawbar to a free standing trailer. This drawbar causes the front axle of the trailer to steer. A major difference between this vehicle and the semi-trailer rig is that each unit is capable of standing alone and there is no transfer of normal force between the units. Figure 6.11. is a plan of a typical vehicle showing the layout of the steering dolly.

A set of equations of motion can be developed in a manner similar to that employed for the semi-trailer by considering the steering dolly as an intermediate body while the body of the trailer is the third body. In practice the steering dolly can be assumed to be massless since its weight is negligible, the function of the dolly is then only kinematic.

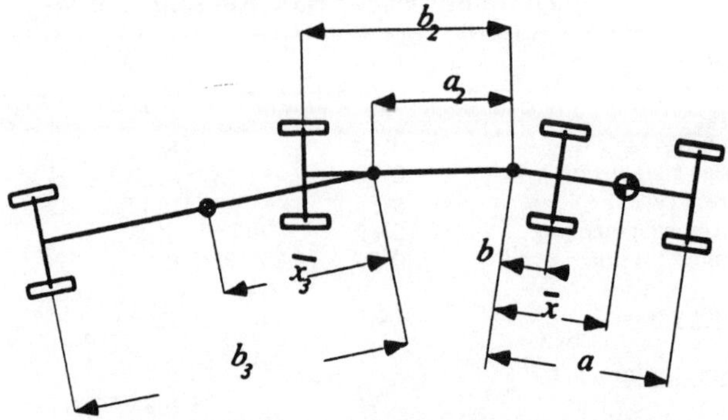

Fig.6.11 The typical layout of a tractor-trailer vehicle

At the pivot between the tractor and steering dolly the equations (6.3) and (6.4) are directly applicable; Fig.6.12(a) shows the relations between the tractor and trailer velocities. Note that the angular velocities, *r*, refer to earth axes. It is now required to develop a further set of kinematic equations relating the rear pivot of the dolly and the steering linkage of the front wheels of the trailer.

Variables are assigned to the dolly and the trailer with appropriate subscripts.

$$U_2 = U\cos\psi + V\sin\psi$$
$$V_2 = -U\sin\psi + V\cos\psi \qquad (6.41)$$

where

$$\psi = \int (r_2 - r)dt$$

Differentiation of equation (6.9) gives the accelerations of the trailer in terms of the tractor variables.

$$\dot{U}_2 = \{\dot{U} + V(r_2 - r)\}\cos\psi + \{\dot{V} - U(r_2 - r)\}\sin\psi$$

$$\dot{V}_2 = -\{\dot{V} - U(r_2 - r)\}\cos\psi + \{\dot{U} + V(r_2 - r)\}\sin\psi$$

Similarly the velocity of the connection between the dolly and the trailer is

$$U_3 = U_2\cos\psi_2 + (V_2 - a_2 r_2)\sin\psi_2$$
$$V_3 = -U_2\sin\psi_2 + (V_2 - a_2 r_2)\cos\psi_2 \qquad (6.42)$$

where

$$\psi_2 = \int (r_3 - r_2)dt$$

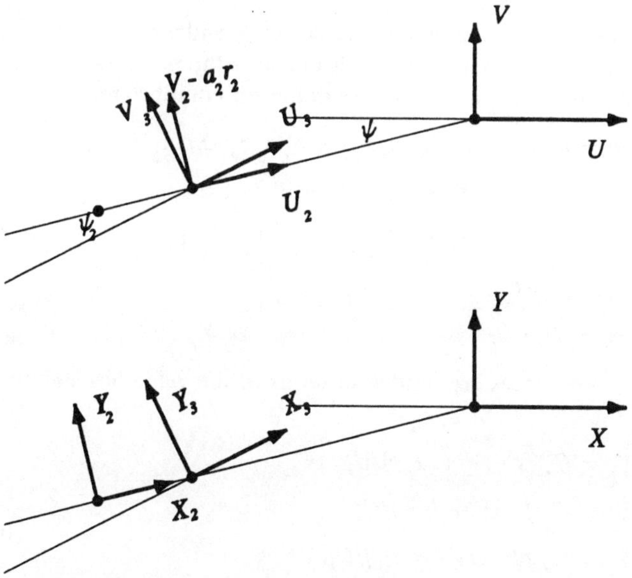

Fig.6.12 (a) The velocities of the front and rear fifth wheels which are related to the steering dolly (b) The forces acting in the steering dolly

Differentiation of equation (6.42) gives the accelerations of the trailer in terms of the tractor variables.

$$\begin{aligned} \dot{U}_3 &= \{\dot{U}_2 + (V_2 - a_2 r_2)(r_3 - r_2)\}\cos\psi_2 \\ &\quad + \{\dot{V}_2 - a_2 \dot{r}_2 - U_2 (r_3 - r_2)\}\sin\psi_2 \\ \dot{V}_3 &= -\{\dot{U}_2 + (V_2 - a_2 r_2)(r_3 - r_2)\}\sin\psi_2 \\ &\quad + \{\dot{V}_2 - a_2 \dot{r}_2 - U_2 (r_3 - r_2)\}\cos\psi_2 \end{aligned} \tag{6.43}$$

It is assumed that the mass and inertia of the steering dolly are small compared with those of the tractor and trailer, hence the accelerations of the dolly are not required and are eliminated.

For small angles of articulation and constant forward speed the lateral acceleration of the trailer is

$$\dot{V}_3 + U_3 r_3 = \dot{V} + Ur - a_2 \dot{r}_2 \tag{6.44}$$

The equations of motion are derived by considering the lateral and yaw equilibrium of each part of the vehicle. For the massless steering dolly summing the moments is the sole condition for equilibrium.

$$\begin{aligned}
&(m_1+m_3)(\dot{V}+Ur) + m_1 \bar{x}_1 \dot{r} - m_3 \bar{a}_2 \dot{r}_2 - m_3 \bar{x}_2 \dot{r}_3 \\
&\qquad = Y_1 + Y_2 + Y_3 + Y_4 \\
&a_2 m_3 (\dot{V}+Ur - a_2 \dot{r}_2 - \bar{x}_3 \dot{r}_3) = a_2 Y_4 - b_2 Y_3 \\
&m_1 x_1 (\dot{V}+Ur) + {}_1I_z \dot{r} = a_1 Y_1 + b_1 Y_2 \\
&-m_3 x_3 (\dot{V}+Ur - a_2 \dot{r}_2) + {}_2I_z \dot{r}_3 = -b_3 Y_4
\end{aligned} \tag{6.45}$$

The lateral tyre forces are written in terms of the velocities and articulation angles.

$$\begin{aligned}
Y_1 &= C_1 \{s - (V + a_1 r)/U\} \\
Y_2 &= C_2 \{-(V + b_1 r)/U\} \\
Y_3 &= C_3 \{\psi_1 - (V - b_2 r_2)/U\} \\
Y_4 &= C_4 \{\psi_1 + \psi_2 - (V - b_3 r_3)/U\}
\end{aligned} \tag{6.46}$$

A steady state analysis of the constant radius, increasing speed responses is developed for three different loading conditions on the trailer. Details of the vehicle are given below. The steer and sidelip angles of the tractor and the articulation angles of the steering dolly and trailer are shown in Fig.6.13 for a typical range of lateral accelerations.

m_1 = 16 000 kg; a_1 = 6.75 m; b_1 = 1.75 m; x_1 = 3.625 m

a_2 = 2.9 m; b_2 = 2.95 m

m_3 = 16 000 kg; b_3 = 4.65 m; x_3 = ?

C_1 = 255 kN/rad; C_2 = 460 kN/rad

C_3 = 318 kN/rad; C_2 = 318 kN/rad

The full trailer vehicle behaves as a linked pair of rigid two axle vehicles. The independence of the tractor is clearly shown in Fig.6.13

which gives the steer and sideslip angles of the tractor and the angles of articulation of the steering dolly and the trailer for a typical vehicle on a 100 metre radius curve. Three centre of mass positions for the trailer are examined.

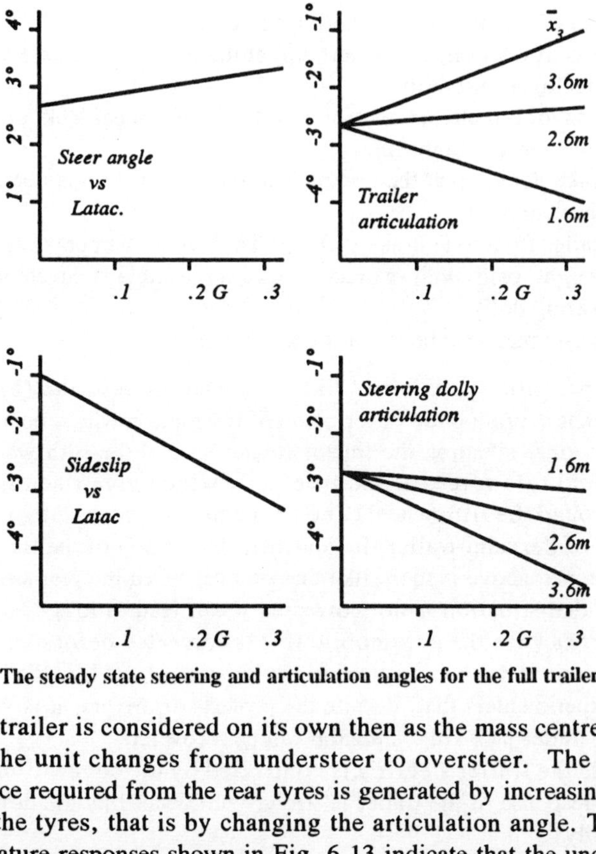

Fig.6.13 The steady state steering and articulation angles for the full trailer vehicle

If the trailer is considered on its own then as the mass centre moves to the rear the unit changes from understeer to oversteer. The increased lateral force required from the rear tyres is generated by increasing the steer angle of the tyres, that is by changing the articulation angle. The steady state curvature responses shown in Fig. 6.13 indicate that the understeering trailer runs with the front end pointing out of the curve and the rear axle cut in. The over-steering trailer has the front axle pointing into the curve and the rear axle running wide. The articulation angles shown are the angles between adjacent units, the angle of the trailer relative to the tractor is the sum of the angles given.

During transient, and particularly reversed, steering the trailer will take up large attitude angles which overshoot the heading angle of the tractor with the result that it can occupy several traffic lanes on a highway. This 'fish-tailing' may also occur as a limit cycle motion when the tractor is proceeding in a straight line. The typical loss of control and unstable situations of the full trailer vehicle are:

(a) Limit cycle yawing oscillation of the trailer.
(b) Loss of traction, wheel spin up, at the tractor drive axle leading to directional instability.
(c) Loss of directional control with front axle brake lock up. Vehicle travels in a straight line.
(d) Brake lock up at the tractor rear axle with the same result as for wheel spin up.
(e) Trailer front axle brake lock up. The trailer attempts to move on a straight path with the result that a large angle is developed at the steering dolly.
(f) Trailer rear axle brake lock up. Trailer swing.

The expression 'jack knife' has been used to describe (b) and (d); however, that is not a good description of the motion which occurs. In the classic jack knife situation the tractor rotates around the fifth wheel driven by the lateral tyre forces at the front axle which give rise to a yawing moment around the fifth wheel. The fifth wheel remains stationary due to the inertia of the semi-trailer. In contrast, the motion of the tractor in the cases mentioned above is more like that of a car when the rear wheels lock, or spin. If that situation is not corrected it can lead to loss of directional control but as with the automobile if it is corrected before large angles occur then it need not be catastrophic. On the other hand it is the experience of many experimenters that, despite the protests of drivers, it is very rarely that a semi-trailer jack knife situation can be recovered.

Because the trailer steering is controlled by the connection with the tractor the response of the trailer is strongly influenced by the design of the steering dolly.

6.25 MULTI-BODY VEHICLE TRAINS

As the number of units in a vehicle train increases the complexity of the analytical model grows and the algebra required to give an explicit set of equations becomes tedious. When a linear version of the analysis is sufficient that is simpler, and in many cases a linear model can be manipulated to provide satisfactory results. A model capable of providing for the analysis of the lateral stability of a double bottom vehicle which consists of a semi-trailer hauling a small full trailer has been published.

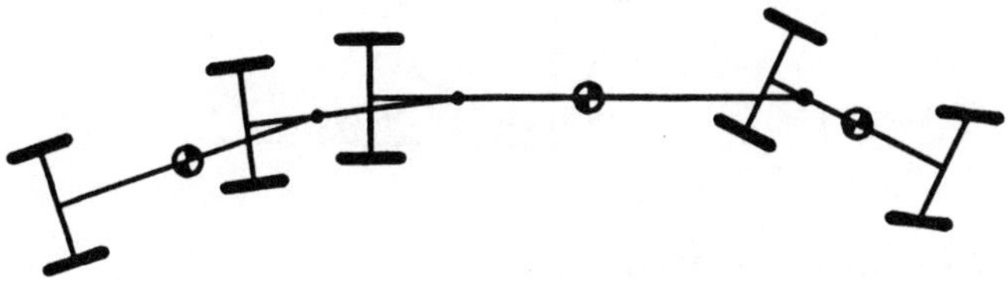

Fig.6.14 An example of a multi-body vehicle

By using the states variable technique described in the development of the four degree of freedom semi-trailer model a non linear simulation can be developed for any vehicle train simply by repeating the equations of motion, the coupling kinematics and force tranfer conditions a sufficient number of times. In this model the origin for each body is taken as the centre of gravity of that body and the fifth wheel velocities given reflect that decision.

Tractor equations:

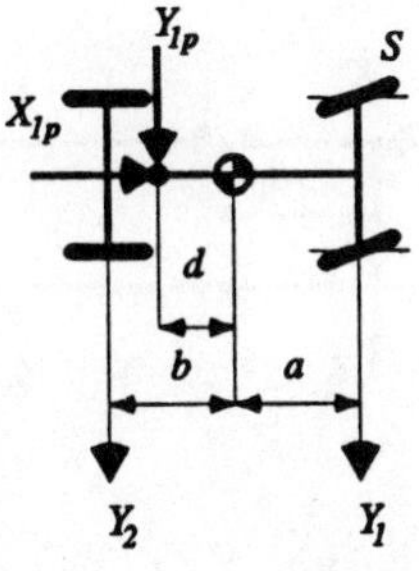

$$m_1(\dot{U} - Vr) = X_{1p} + X_1 + X_2$$

$$m_1(\dot{V} + Ur) = -Y_{1p} + Y_1 + Y_2$$

$$I_1\dot{r} = -dY_{1p} + aY_1 - bY_2$$

$$Y_1 = f\{s - (V + ar)/U\}$$

$$Y_2 = f\{-(V - br)/U\}$$

Equate the velocities at the first fifth wheel

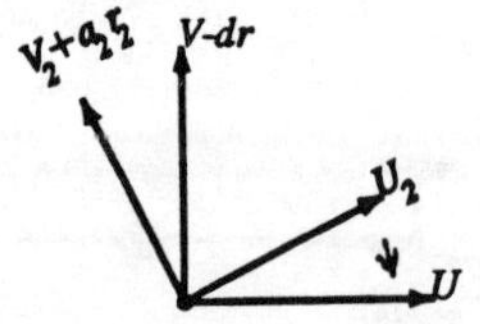

$$U_2 = U\cos\psi + (V - dr)\sin\psi$$
$$V_2 = -U\sin\psi + (V - dr)\cos\psi - a_2 r_2$$

Hence

$$\dot{U}_2 = \{\dot{U} - (V - dr)(r_2 - r)\}\cos\psi + \{\dot{V} - d\dot{r} - U(r_2 - r)\}\sin\psi$$
$$\dot{V}_2 = -\{\dot{U} - (V - dr)(r_2 - r)\}\sin\psi + [\dot{V} - d\dot{r} - U(r_2 - r)\}\cos\psi \quad (6.47)$$
$$\psi = \int (r_2 - r)\cdot dt$$

First semi-trailer equations:

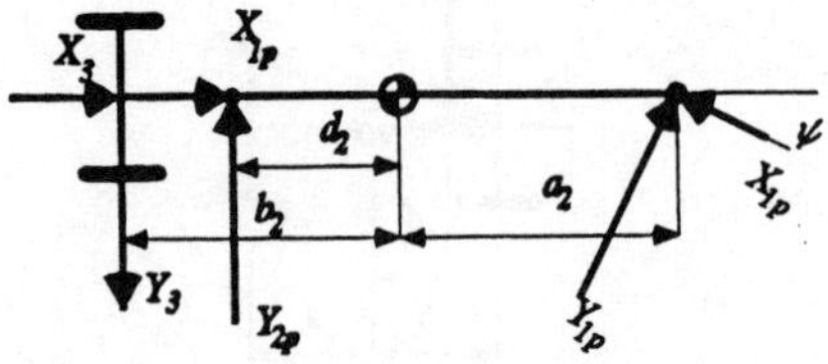

$$m_2(\dot{U}_2 - V_2 r_2) = X_{2p} - (X_{1p}\cos\psi + Y_{1p}\sin\psi) + X_3$$
$$m_2(\dot{V}_2 - u_2 r_2) = Y_{2p} - (X_{1p}\sin\psi - Y_{1p}\cos\psi) + Y_3$$
$$I_2\dot{r} = -d_2 Y_{2p} + a_2(X_{1p}\sin\psi - Y_{1p}\cos\psi) - b_2 Y_3$$
$$Y_3 = f\{\psi - (V_2 - b_2 r_2)/U_2\}$$

At the connection of steering dolly to semi-trailer the velocities are equated and the accelerations are deduced.

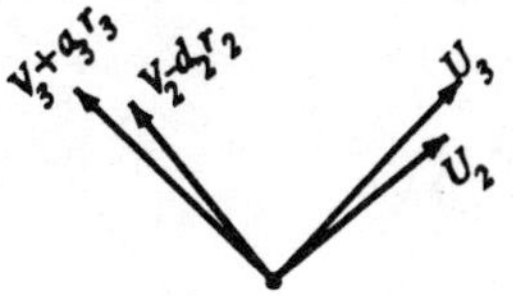

Second fifth wheel:

$$U_3 = U_2 \cos\psi_2 + (V_2 - d_2 r_2)\sin\psi_2$$
$$V_3 = - U_2 \sin\psi_2 + (V_2 - d_2 r_2)\cos\psi_2 - a_3 r_3$$

Hence

$$U_3 = \{\dot{U}_2 - (V_2 - d_2 r_2)(r_3 - r_2)\}\cos\psi_2$$
$$+ \{\dot{V}_2 - d_2 \dot{r}_2 - U_2 (r_3 - r_2)\}\sin\psi_2$$
$$V_2 = - \{\dot{U}_2 - (V_2 - d_2 r_2)(r_3 - r_2)\}\sin\psi_2 \quad (6.48)$$
$$+ \{\dot{V}_2 - d_2 \dot{r}_2 - U_2 (r_3 - r_2)\}\cos\psi_2$$
$$\psi_2 = \int (r_3 - r_2)\cdot dt$$

The mass and inertia of the steering dolly are neglected. Thus the equations for the steering dolly are the force and moment balance.

Steering dolly equations:

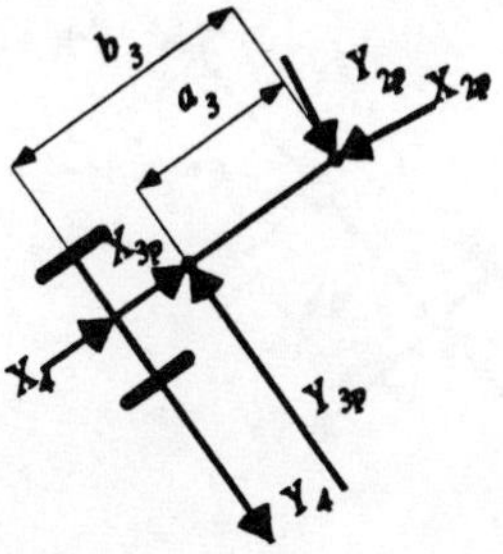

$$0 = - X_{3p} + X_{2p} \cos\psi_2 + Y_{2p} \sin\psi_2$$
$$0 = - Y_{3p} + Y_{2p} \cos\psi_2 - X_{2p} \sin\psi_2$$
$$0 = d_3 Y_{3p} + b_3 Y_4$$
$$Y_4 = f\{\psi + \psi_2 - (V_3 - b_3 r_3)/U_3\}$$

Third fifth wheel:

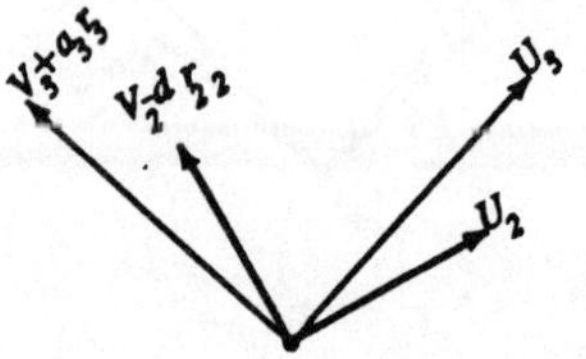

$$U_4 = U_3 \cos\psi_3 + (V_3 - d_3 r_3)\sin\psi_3$$
$$V_4 = -U_3 \sin\psi_3 + (V_3 - d_3 r_3)\cos\psi_3 - a_4 r_4$$

Hence

$$\dot{U}_4 = \{\dot{U}_3 - (V_3 - d_3 r_3)(r_4 - r_3)\}\cos\psi_3$$
$$+ \{\dot{V}_3 - d_3 \dot{r}_3 - U_3 (r_4 - r_3)\}\sin\psi_3$$
$$\dot{V}_4 = -\{U_3 - (\dot{V}_3 - d_3 r_3)(r_4 - r_3)\}\sin\psi_3$$
$$+ \{\dot{V}_3 - \dot{d}_3 r_3 - U_3 (r_4 - r_3)\}\cos\psi_3$$

Second trailer:

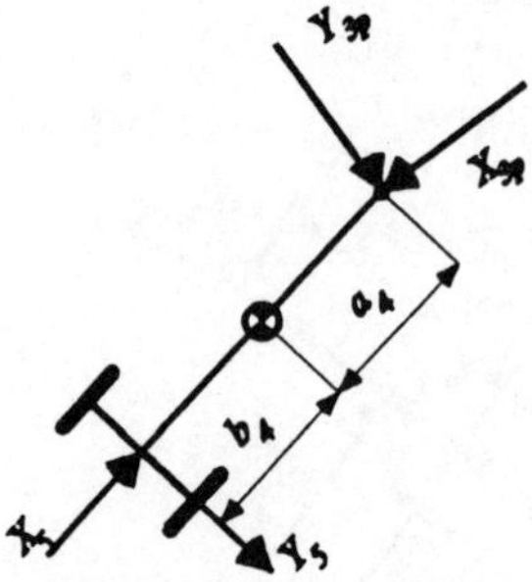

$$m_4(\dot{U}_4 - V_4 r_4) = -(X_{3p}\cos\psi_3 + Y_{3p}\sin\psi_3) + X_5$$
$$m_4(\dot{V}_4 - u_4 r_4) = -(X_{3p}\sin\psi_3 - Y_{3p}\cos\psi_3) + Y_5$$
$$I_4 \dot{r} = a_4(X_{3p}\sin\psi_3 - Y_{3p}\cos\psi_3) - b_4 Y_5$$
$$Y_5 = f\{\psi + \psi_2 + \psi_3 - (V_4 - b_4 r_4)/U_4\}$$

The braking and tractive instabilities of the extended semi-trailer vehicle are similar to those of the basic semi-trailer with the addition of two modes from the additional unit.

Locking the front axle induces total stability and the path becomes tangent to the curve while locking the tractor rear axle causes the typical jack-knife instability. When the wheels of the first trailer are locked the resulting outward swing of that trailer steers the dolly supporting the front of the second trailer while the wheels of that unit and the second trailer develop forces which attempt to re-align the second trailer. The steering dolly swings out when the wheels are locked with the result that the following trailer is shifted off track but maintains its directional heading. Finally the second trailer swings outward as the final axle of the vehicle train is locked.

6.26 TRIM ANALYSIS

It is often desirable to study the characteristics of a vehicle under conditions other than those around the straight ahead. One way of doing this is to run a simulation which can be taken through the required manoeuver and the derivatives of tyre forces, spring and damper recorded. These data are then applied to linearised models to give the changes in the responses from those of the straight running vehicle.

When a complex simulation is not available, or when the data needed cannot be obtained without difficulty then the methods described in sections 6.5 and 6.6 enable the steady state tyre characteristics to be computed at any lateral acceleration and braking or traction condition. Thus a trim analysis can be carried out for various steady state lateral accelerations.

6.27 ARTICULATED STEER VEHICLES

Vehicles which steer by rotating two bodies, each of which carries an axle, around a central post are more frequent in off road operations than on the highway. This type of vehicle may be studied by considering it as an articulated semi trailer with the front axle removed. The control angle is generated by relative rotation of the two parts.

The steering system is usually a set of hydraulic rams and a linkage. This will be represented by a rotational displacement with a spring interposed between the rams and the vehicle to permit a degree of flexibility. Because there is a natural tendency for the vehicle to collapse around the hinge the steering is active at all times and the stiffness is important. The vehicle may be subject to static instability when it is stationary and the two parts are not in line. Further instability may be induced by the steering gear itself. For example when backlash is present it has been noted that turning the steering wheel with the vehicle stationary and then releasing it may result in a limit cycle oscillation in which the two parts of the vehicle oscillate in the yaw mode without damping. This cycle can be halted by braking the road wheels.

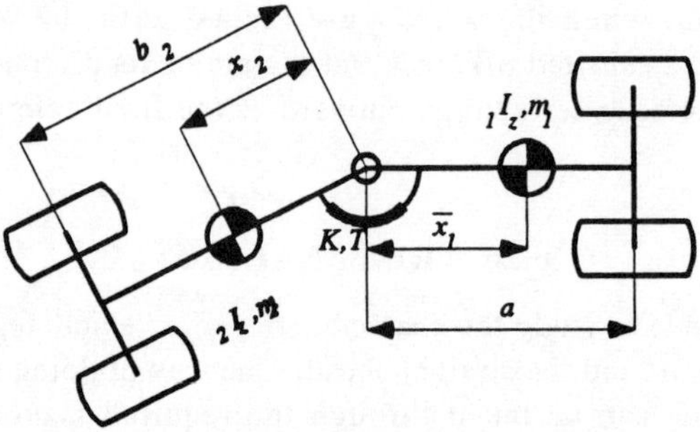

Fig.6.15 The articulated steer vehicle is controlled by changing the angle between the two parts. The model is equivalent to a semi-trailer vehicle with a single axle on the front body. The steering system controls the angle between the bodies, the stiffness of this unit is important

The most basic model is that originally derived in equation 6.17 which is now modified to represent the elementary steering system.

$$m(\dot{V}+Ur) + m_1\bar{x}_1\dot{r} - m_2\bar{x}_2\dot{r}_2 = Y_f + Y_t$$

$$m_1\bar{x}_1(\dot{V}+Ur) + {}_1I_z\dot{r} = aY_f + K(S - \psi)$$

$$-m_2\bar{x}_2(\dot{V}+Ur) + {}_2I_z\dot{r}_2 = -b_2Y_t - K(S - \psi)$$

The linearised tyre forces are written in terms of these variables with the assumption of a common attitude angle for both the sets of tyres on a single axle.

Front body: $Y_f = C_f\{ - (V + ar)/U\}$

Rear body: $Y_t = C_t\{\psi - (V - b_2r_2)\}/U$

$$\begin{bmatrix} m & m_1\bar{x}_1 & -m_2\bar{x}_2 \\ m_1\bar{x}_1 & {}_1I_z & 0 \\ -m_2\bar{x}_2 & 0 & {}_2I_z \end{bmatrix} \begin{bmatrix} V \\ r \\ r_2 \end{bmatrix} =$$

$$\begin{bmatrix} -(C_f+C_t)/U & -mU-(aC_f)/U & b_2C_t/U & C_t & 0 \\ -aC_f/U & -m_1\bar{x}_1U-(a^2C_f)/U & 0 & -K & K \\ b_2C_t/U & m_2\bar{x}_2U & -b_2{}^2C_t/U & K-b_2C_t & -K \end{bmatrix} \begin{bmatrix} V \\ r \\ r_2 \\ \psi \\ s \end{bmatrix}$$

Matrix 6.6 The vehicle with steering by frame articulation

In the steady state the value of r_2 is the same as r, thus the two values are combined.

$$\begin{bmatrix} C_f+C_t & mU+(aC_f-b_2C_t)/U & -C_t \\ aC_f & m_{\bar{1}1}U+a^2C_f/U & K \\ -b_2C_t & -m_2\bar{x}_2U+b_2^2C_t/U & -K+b_2C_t \end{bmatrix} \begin{bmatrix} \beta \\ r \\ \psi \end{bmatrix} = \begin{bmatrix} 0 \\ K \\ -K \end{bmatrix} S$$

Matrix 6.7 The steady state steering response equations for the linear articulated steering vehicle

The steady state responses depend upon the fact that the characteristic determinant is not zero.

$$\begin{vmatrix} C_f+C_t & mU+(aC_f-b_2C_t)/U & -C_t \\ aC_f & m_1\bar{x}_1U+a^2C_f/U & K \\ -b_2C_t & -m_2\bar{x}_2U+b_2^2C_t/U & -K+b_2C_t \end{vmatrix} \neq 0$$

Sideslip response

$$\beta/S_{ss} = \frac{\begin{vmatrix} 0 & mU+(aC_f-b_2C_t)/U & -C_t \\ K & m_1\bar{x}_1U+a^2C_f/U & K \\ -K & -m_2\bar{x}_2U+b_2^2C_t/U & -K+b_2C_t \end{vmatrix}}{\begin{vmatrix} C_f+C_t & mU+(aC_f-b_2C_t)/U & -C_t \\ aC_f & m_1x_1U+a^2C_f/U & K \\ -b_2C_t & -m_2x_2U+b_2^2C_t/U & -K+b_2C_t \end{vmatrix}}$$

Yaw velocity response:

$$r/S_{ss} = \frac{\begin{vmatrix} C_f+C_t & 0 & -C_t \\ aC_f & K & K \\ -b_2C_t & -K & -K+b_2C_t \end{vmatrix}}{\begin{vmatrix} C_f+C_t & mU+(aC_f-b_2C_t)/U & -C_t \\ aC_f & m_1\bar{x}_1U+a^2C_f/U & K \\ -b_2C_t & -m_2\bar{x}_2U+b_2^2C_t/U & -K+b_2C_t \end{vmatrix}}$$

Articulation angle

$$\psi/S_{ss} = \frac{\begin{vmatrix} C_f+C_t & mU+(aC_f-b_2C_t)/U & 0 \\ aC_f & m_1\bar{x}_1U+a^2C_f/U & K \\ -b_2C_t & -m_2\bar{x}_2U+b_2^2C_t/U & -K \end{vmatrix}}{\begin{vmatrix} C_f+C_t & mU+(aC_f-b_2C_t)/U & -C_t \\ aC_f & m_1\bar{x}_1U+a^2C_f/U & K \\ -b_2C_t & -m_2\bar{x}_2U+b_2^2C_t/U & -K+b_2C_t \end{vmatrix}}$$

6.28 ARTICULATED STEER WITH FLEXIBLE TYRES

The model derived above is based on a 'slip angle' tyre model and is thus unable to represent the non-rolling and slow speed, forward and reverse operation of the articulated frame steer vehicle. However it is possible to use the flexible tyre model developed in chapter 1 to show tyre steady state and transient behaviour and also applied in chapter 4, for a more universal model. This model will have the same steady state responses as the previous example but is not restricted by the need for forward motion. The equations of motion are derived from the previous example. The tyres are assumed to be similar for this case and a single value of lateral stiffness, K', is used.

$$m(\dot{V}+Ur) + m_1\bar{x}_1\dot{r} - m_2\bar{x}_2\dot{r}_2 = K'y_f + K'y_t$$

$$m_1\bar{x}_1(V+Ur) + {}_1I_z\dot{r} = aK'y_f + K(S-\psi)$$

$$-m_2\bar{x}_2(V+Ur) + {}_2I_z\dot{r}_2 = -b_2K'y_t - K(S-\psi)$$

The linearised tyre forces relate lateral deflection to the attitude angle. Note the change in slip angle due to the lateral tyre deflection

Front body: $K'y_f = C_f\{-(V + ar + \dot{y}_f)/U\}$

Rear body: $K'y_t = C_t\{\psi - (V - b_2r_2 + \dot{y}_t\})/U$

The equations of motion are then written in matrix form.

$$\begin{bmatrix} m & m_1x_1 & -m_2x_2 & 0 & 0 \\ m_1x_1 & {}_1I_z & 0 & 0 & 0 \\ -m_2x_2 & 0 & {}_2I_z & 0 & 0 \\ 0 & 0 & 0 & 1 & 0 \\ 0 & 0 & 0 & 0 & 1 \end{bmatrix} \begin{bmatrix} V \\ r \\ r_2 \\ y_f \\ y_r \end{bmatrix} = \begin{bmatrix} 0 & -mU & 0 & 0 & 0 & 0 \\ 0 & m_1x_1 & -K & 0 & 0 & K \\ 0 & -m_2x_2 & K & 0 & 0 & -K \\ -1 & -a & U & 1 & 0 & 0 \\ -1 & b_2 & 0 & 0 & 1 & 0 \end{bmatrix} \begin{bmatrix} V \\ r \\ \psi \\ y_f \\ y_t \\ S \end{bmatrix}$$

Matrix 6.8 The equations of motion for an articulated steering vehicle on flexible tyres.

This model can be extended to include longitudinal tyre characteristics if desired, for many purposes it is sufficient to include only a term for the tyre rolling on the ground.

6.29 CONCLUSIONS

This chapter has traced the development of handling models for articulated vehicles and shown some of the design problems that can be solved with their use. It has not attempted to introduce very complex simulations which may be tailored to fit every possible design variation n but rather takes the position that basic models which provide readily understood results can be modified for particular problems. These basic models for the semi-trailer vehicle, twin tyres and multiple axles are derived.

REFERENCES

(1) E.F.KURTZ AND R.J.ANDERSON (1977) Handling characteristics of car-trailer systems, a state-of-the-art survey, *Vehicle System Dynamics*, **6**, 217-243.

(2) J.R.ELLIS (1977) The articulated semi-trailer vehicle including the roll mode, Conference on Vehicle System Dynamics, Vienna.

(3) F.VLK (1985) Handling performance of truck-trailer vehicles: a state-of-the-art survey, *Int. J. Vehicle Design*, **6**(3).

Index